AF779909

Thin Film Microelectronic Devices and Circuits

Thin Film Microelectronic Devices and Circuits

Guest Editor

Chengyuan Dong

MDPI

Basel • Beijing • Wuhan • Barcelona • Belgrade • Novi Sad • Cluj • Manchester

Guest Editor
Chengyuan Dong
Department of Electronic
Engineering
Shanghai Jiao Tong University
Shanghai
China

Editorial Office
MDPI AG
Grosspeteranlage 5
4052 Basel, Switzerland

This is a reprint of the Special Issue, published open access by the journal *Micromachines* (ISSN 2072-666X), freely accessible at: www.mdpi.com/journal/micromachines/special_issues/Y39J6Z5142.

For citation purposes, cite each article independently as indicated on the article page online and using the guide below:

Lastname, A.A.; Lastname, B.B. Article Title. *Journal Name* **Year**, *Volume Number*, Page Range.

ISBN 978-3-7258-3300-9 (Hbk)
ISBN 978-3-7258-3299-6 (PDF)
https://doi.org/10.3390/books978-3-7258-3299-6

© 2025 by the authors. Articles in this book are Open Access and distributed under the Creative Commons Attribution (CC BY) license. The book as a whole is distributed by MDPI under the terms and conditions of the Creative Commons Attribution-NonCommercial-NoDerivs (CC BY-NC-ND) license (https://creativecommons.org/licenses/by-nc-nd/4.0/).

Contents

About the Editor

Chengyuan Dong

Chenyuan Dong received his B.S. and Ph.D. from Shanghai Jiao Tong University in 1993 and 2003, respectively. From 2003 to 2008, he worked as a principal engineer/project manager in China's main TFT-LCD makers including SVA-NEC, IVO, and Tianma Microelectronics. Dr. Dong, as the key member, joined the setup of the first G5 TFT-LCD fabrication line in China. From Nov. 2008, Dr. Dong has been an associate professor at Shanghai Jiao Tong University, where he has been focusing on teaching and research in the field of thin-film electronic devices and integrations. Up to now, he has published one book, more than 60 papers, and 10 Chinese patents.

micromachines

Editorial

Editorial for the Special Issue on Thin Film Microelectronic Devices and Circuits

Chengyuan Dong

Department of Electronic Engineering, Shanghai Jiao Tong University, Shanghai 200240, China; cydong@sjtu.edu.cn

Thin film microelectronic devices and circuits (TFMDCs), including thin film transistors (TFTs), thin film solar cells (TFSCs), thin film sensors (TFSs), thin film memories (TFMs), etc., are playing more and more important roles in electronic industries, such as integrated circuits, semiconductor displays, medical devices, energy devices, optical sensors, and so on [1–3]. One of the outstanding feathers of TFMMDCs is flexibility, which results in some brand-new products and applications, such as flexible displays and memories [4–6], wearable cells and sensors [7,8], and stretchable medical devices and systems [9,10]. On the other hand, some TFMDCs with special merits can penetrate into the conventional industries and push them forward. For example, amorphous oxide semiconductor (AOS) TFTs show extraordinarily low leakage currents, and can be combined with the back-end of line (BEOL) of complementary metal-oxide-semiconductor (CMOS) integrated circuits, which benefit their applications in low-power dynamic random access memories (DRAMs) [4]. To realize these interesting applications, TFMDCs are still crying for advancements in fabrication technologies, device physics, circuit designs, and system integrations.

This Special Issue comprises 10 original papers about recent advances in the research and development of TFMDCs. Specifically, three microelectronic devices are covered here: thin film transistors (three papers), two-terminal thin film components (five papers), and optical devices (two papers). These typical studies exhibit the recent advances in this interesting field, which are briefly summarized as follows.

For AOS-TFTs, amorphous InGaZnO (a-IGZO) films are the most popular channel layers, which have already been put into mass production [11]. Therefore, many attempts have been taken to improve their electrical performance and stable properties. In this Special Issue, H. Huang et al. [12] investigate the influence of hydrogen buffer layer on the electrical performances of top-gate IGZO-TFTs, indicating that the hydrogen content of the buffer layer increases proportionally with the rise in the hydrogen content of the reaction gases during plasma-enhanced chemical vapor deposition (PECVD); in addition, with the increase in the hydrogen-containing materials in the reactive gases, the field-effect mobility and negative bias illumination stress (NBIS) stability of the corresponding IGZO-TFTs improve nearly twofold.

Gate insulators also exhibit strong effects on the electrical properties of IGZO-TFTs [13]; high-k metal oxides are gradually replacing the traditional SiO_2 dielectric layer in the new generation of microelectronic devices. In this Special Issue, J. Liu et al. [14] report the production of five-element high entropy metal oxide (HEMO) dielectric films by the solution method, analyzing the role of each metal oxide in the system by characterizing the film's properties. The IGZO-TFTs with $(AlGaTiYZr)O_x$ dielectric layers show a mobility of 182 cm^2/Vs, a threshold voltage of -0.2 V, and a subthreshold swing of 0.3 V/dec, which implies a good prospect for applying HEMOs to TFT devices.

check for updates

Received: 24 January 2025
Accepted: 28 January 2025
Published: 30 January 2025

Citation: Dong, C. Editorial for the Special Issue on Thin Film Microelectronic Devices and Circuits. *Micromachines* **2025**, *16*, 167. https://doi.org/10.3390/mi16020167

Copyright: © 2025 by the author. Licensee MDPI, Basel, Switzerland. This article is an open access article distributed under the terms and conditions of the Creative Commons Attribution (CC BY) license (https://creativecommons.org/licenses/by/4.0/).

In addition to a-IGZO, the other AOS channel layers for thin film transistors have also been being studied [15]. In this Special Issue, Z. Wu et al. [16] performed an annealing study on the praseodymium-doped indium zinc oxide (PrIZO) TFTs, which indicates that the channel films tend to be denser and obtain a lower surface roughness, a narrower optical-band gap, and less oxygen-vacancy defects for the higher annealing temperatures; additionally, the PrIZO-TFTs annealed at 150 °C exhibit a desired performance, as well as good flexible properties.

The solution-processed dielectric films can be used as not only the gate insulators of TFT devices, but also the core layers of metal–insulator–metal (MIM) and metal–insulator–semiconductor (MIS) capacitors [17–19]. In this Special Issue, X. Fang et al. [20] utilized a low-temperature self-exothermic reaction based on the solution method to prepare high-performance Al_2O_3 dielectric thin films on the flexible substrates; the corresponding MIM devices demonstrate excellent electrical performances, including a leakage current density as low as 1.08×10^{-8} A/cm^2 @1 MV and a relative dielectric constant as high as 8.61 ± 0.06.

There is another report relating MIM capacitors in this Special Issue. T. M. Choi et al. [21] investigated the MIM capacitors with different thicknesses of SixNy film and varying levels of film quality to improve their capacitance density. In this study, the C-V characteristics are divided into two categories: the voltage coefficient of capacitance (VCC) and the temperature coefficient of capacitance (TCC). When the thickness of the SixNy film decreases, the VCC increases, whereas the TCC does not vary significantly; the most influential factor for capacitance uniformity is the thickness uniformity of the SixNy film.

The novel dielectric films for MIS capacitors are also reported in this Special Issue. J. U. Yoo et al. [22] proposed a new method to effectively fabricate a poly(vinylidene fluoride) (PVDF)-based TiO_2 dielectric layer via electrospinning. Improved electrical properties are observed with increasing TiO_2 anatase content, and the residual amount of PVDF decreased with increasing annealing temperature; the corresponding MIS capacitors annealed at 600 °C exhibit a lower leakage current density of 7.5×10^{-13} A/cm^2 when Vg = 0 V.

Memristor is one of the most interesting two-terminal thin film devices, which might be put into wide applications in the future [23]. However, more systematic studies on the applications of memristors are still necessary at the present stage. In this Special Issue, Y. Kim et al. [24] employed a study of weight quantization associations over a weight range for application in memristors. To minimize the information corruption in the system caused by weight quantization, the concept of "weight range" was introduced, which has a direct impact on weight, reducing the number of digits represented by a weight below a certain level.

MoS_2 films have been attracting much attention due to their interesting properties [25]. There is an interesting study about MoS_2 and its device by pulsed laser deposition in this Special Issue [26]. The authors gradually increased the pulsed laser energy density from 70 mJ·cm^{-2}, and finally prove that 100 mJ·cm^{-2} is the best-pulsed laser energy density. The Si/MoS_2 heterojunction prepared under the optimized laser energy density indicates an opening voltage of 0.61 V and reaction ratio of 457.0.

Lately, thin film optical devices are playing more and more important roles in the actual applications [27,28]. In this Special Issue, two reports regarding interferometers and light-detection-sensors are presented. J. Liao et al. [29] propose a polarization-insensitive lithium niobate-on-insulator (LN) interferometer, and make further improvements. By elaborately designing the geometric structure of the multimode interference coupler, beam splitting of equal proportions and directional coupling of higher-order modes are realized. At 1550 nm, the visibility of the interferometer is 97.59% and 98.16% for transverse electric (TE) and transverse magnetic (TM) modes, respectively, demonstrating the high performance of the proposed LN polarization-independent interferometer.

Traditional light-direction angle sensors relying on optical components like gratings and lenses face inherent constraints form diffraction limits in achieving device miniaturization. In this Special Issue, P. Huang et al. [30] proposed a sensor concept capable of discerning the 3D light-direction based on a segmented concentric nanoring structure, which is sensitive to both elevation angle and azimuth angle at a micrometer device scale. The concept was validated through simulation studies. This design broadens the angle sensing dimension based on mutual resonance coupling among nanoring segments, and through waveguide implementation or sensor array arrangements. The detection range can be flexibly adjusted to accommodate diverse application scenarios.

Finally, I would like to thank all of the authors for submitting their papers to the Special Issue "Thin Film Microelectronic Devices and Circuits", as well as all the reviewers and editors for their contributions to improving these submissions.

Conflicts of Interest: The author declares no conflicts of interest.

References

1. Kasap, S.; Capper, P. *Handbook of Electronic and Photonic Materials*; Springer International Publishing: Cham, Switzerland, 2017.
2. Hosono, H.; Paine, D.C. *Handbook of Transparent Conductors*; Springer International Publishing: Cham, Switzerland, 2010.
3. Ramanathan, S. *Thin Film Metal-Oxides: Fundamentals and Applications in Electronics and Energy*; Springer International Publishing: Cham, Switzerland, 2010.
4. Hosono, H.; Kumomi, H. *Amorphous Oxide Semiconductors: IGZO and Related Materials for Display and Memory*; Wiley-VCH Verlag & Co. KgaA: Hoboken, NJ, USA, 2022.
5. Brotherton, S.D. *Introduction to Thin Film Transistors*; Springer International Publishing: Cham, Switzerland, 2013.
6. Zhou, Y. *Semiconducting Metal Oxide Thin-Film Transistors*; IOP Publishing Ltd.: Bristol, UK, 2021.
7. Khanna, V.K. *Flexible Electronics: Energy Devices and Applications*; IOP Publishing: Bristol, UK, 2019.
8. Mukhopadhyay, S.C. *Wearable Electronics Sensors*; Springer International Publishing: Cham, Switzerland, 2015.
9. Someya, T. *Stretchable Electronics*; Wiley-VCH Verlag & Co. KgaA: Hoboken, NJ, USA, 2013.
10. Rogers, J.A.; Ghaffari, B.; Kim, D.H. *Stretchable Bioelectronics for Medical Devices and Systems*; Springer International Publishing: Cham, Switzerland, 2016.
11. Park, J.C.; Ahn, S.-E.; Lee, H.-N. High-Performance Low-Cost Back-Channel-Etch Amorphous Gallium–Indium–Zinc Oxide Thin-Film Transistors by Curing and Passivation of the Damaged Back Channel. ACS Appl. *Mater. Interfaces* **2013**, *5*, 12262. [CrossRef] [PubMed]
12. Huang, H.; Peng, C.; Xu, M.; Chen, L.; Li, X. Dependence of a Hydrogen Buffer Layer on the Properties of Top-Gate IGZO TFT. *Micromachines* **2024**, *15*, 722. [CrossRef] [PubMed]
13. Robertson, J. High dielectric constant gate oxides for metal oxide Si transistors. *Rep. Prog. Phys.* **2006**, *69*, 327. [CrossRef]
14. Liu, J.; Xiong, X.; Li, H.; Huang, X.; Wang, Y.; Sheng, Y.; Liang, Z.; Yao, R.; Ning, H.; Wei, X. Application of Solution-Processed High-Entropy Metal Oxide Dielectric Layers with High Dielectric Constant and Wide Bandgap in Thin-Film Transistors. *Micromachines* **2024**, *15*, 1465. [CrossRef] [PubMed]
15. Xu, W.Y.; Li, H.; Xu, J.B.; Wang, L. Recent Advances of Solution-Processed Metal Oxide Thin-Film Transistors. *ACS Appl. Mater. Interfaces* **2018**, *10*, 25878. [CrossRef]
16. Wu, Z.; Ning, H.; Li, H.; Wei, X.; Luo, D.; Yuan, D.; Liang, Z.; Su, G.; Yao, R.; Peng, J. Annealing Study on Praseodymium-Doped Indium Zinc Oxide Thin-Film Transistors and Fabrication of Flexible Devices. *Micromachines* **2025**, *16*, 17. [CrossRef] [PubMed]
17. Huan, T.D.; Boggs, S.; Teyssedre, G.; Laurent, C.; Cakmak, M.; Kumar, S.; Ramprasad, R. Advanced Polymeric Dielectrics for High Energy Density Applications. *Prog. Mater. Sci.* **2016**, *83*, 236. [CrossRef]
18. Choi, E.; Kim, A.; Kwon, S.H.; Pyo, S.G. Effect of Interface Treatment on the Voltage Linearity in 8 fF/μm2 High-k Dielectric and Combination Stacks on Metal Insulator Metal (MIM) Capacitor. *Sci. Adv. Mater.* **2018**, *10*, 467–470. [CrossRef]
19. Shubham, K.; Khan, R. Annealing effect on fabrication and characterization of MIS structure using TiO2-SiO2 Thin Film as Insulator layer deposited by low temperature arc vapor deposition process. *J. Electron Devices* **2013**, *17*, 1439.
20. Fang, X.; Ning, H.; Zhang, Z.; Yao, R.; Huang, Y.; Yang, Y.; Cheng, W.; Jin, S.; Luo, D.; Peng, J. Preparation of High-Performance Transparent Al2O3 Dielectric Films via Self-Exothermic Reaction Based on Solution Method and Applications. *Micromachines* **2024**, *15*, 1140. [CrossRef]
21. Choi, T.M.; Jung, E.S.; Yoo, J.U.; Lee, H.R.; Pyo, S.G. Capacitance–Voltage Fluctuation of Si_xN_y-Based Metal–Insulator–Metal Capacitor Due to Silane Surface Treatment. *Micromachines* **2024**, *15*, 1204. [CrossRef] [PubMed]

22. Yoo, J.U.; Choi, T.M.; Pyo, S.G. Fabrication of Electrospun Porous TiO_2 Dielectric Film in a Ti–TiO_2–Si Heterostructure for Metal–Insulator Semiconductor Capacitors. *Micromachines* **2024**, *15*, 1231. [CrossRef] [PubMed]
23. Yao, P.; Wu, H.; Gao, B.; Tang, J.; Zhang, Q.; Zhang, W.; Yang, J.J.; Qian, H. Fully hardware-implemented memristor convolutional neural network. *Nature* **2020**, *577*, 641. [CrossRef] [PubMed]
24. Kim, Y.; Noh, H.Y.; Koo, G.; Lee, H.; Lee, S.; Choi, R.H.; Lee, S.; Lee, M.J.; Lee, H.J. Study of Weight Quantization Associations over a Weight Range for Application in Memristor Devices. *Micromachines* **2024**, *15*, 1258. [CrossRef] [PubMed]
25. Liu, Y.; Ong, Z.-Y.; Wu, J.; Zhao, Y.; Watanabe, K.; Taniguchi, T.; Chi, D.; Zhang, G.; Thong, J.T.L.; Qiu, C.-W.; et al. Thermal Conductance of the 2D MoS2/h-BN and Graphene/h-BN Interfaces. *Sci. Rep.* **2017**, *7*, 43886. [CrossRef]
26. Cai, W.; Liu, Y.; Yao, R.; Yuan, W.; Ning, H.; Huang, Y.; Jin, S.; Fang, X.; Guo, R.; Peng, J. Optimization of Pulsed Laser Energy Density for the Preparation of MoS_2 Film and Its Device by Pulsed Laser Deposition. *Micromachines* **2024**, *15*, 945. [CrossRef] [PubMed]
27. Haurylau, M.; Chen, G.; Chen, H.; Zhang, J.; Nelson, N.A.; Albonesi, D.H.; Friedman, E.G.; Fauchet, P.M. On-Chip Optical Interconnect Roadmap: Challenges and Critical Directions. *IEEE J. Sel. Top. Quantum Electron.* **2006**, *12*, 1699. [CrossRef]
28. Nagarajan, A.; Hara, S.; Satoh, H.; Panchanathan, A.P.; Inokawa, H. Angle-sensitive detector based on silicon-on-insulator photodiode stacked with surface plasmon antenna. *Sensors* **2020**, *20*, 5543. [CrossRef]
29. Liao, J.; Liu, L.; Sun, Y.; Wang, Z.; Li, W.; Lan, J.; Ma, L.; Lu, Z. Polarization-Insensitive Lithium Niobate-on-Insulator Interferometer. *Micromachines* **2024**, *15*, 983. [CrossRef] [PubMed]
30. Huang, P.; Wu, P.; Guo, Z.; Ye, Z. 3DLight-Direction Sensor Based on Segmented Concentric Nanorings Combined with Deep Learning. *Micromachines* **2024**, *15*, 1219. [CrossRef] [PubMed]

Disclaimer/Publisher's Note: The statements, opinions and data contained in all publications are solely those of the individual author(s) and contributor(s) and not of MDPI and/or the editor(s). MDPI and/or the editor(s) disclaim responsibility for any injury to people or property resulting from any ideas, methods, instructions or products referred to in the content.

micromachines

Article

Dependence of a Hydrogen Buffer Layer on the Properties of Top-Gate IGZO TFT

Huixue Huang [1,2], Cong Peng [1,2,*], Meng Xu [1,2], Longlong Chen [2] and Xifeng Li [1,2,*]

1 Shanghai Collaborative Innovation Center of Intelligent Sensing Chip Technology, Shanghai University, Shanghai 201800, China; hhx0519@163.com (H.H.); xumeng@shu.edu.cn (M.X.)
2 Key Laboratory of Advanced Display and System Applications of Ministry of Education, Shanghai University, Shanghai 200072, China; llchen@i.shu.edu.cn
* Correspondence: pengcong92@163.com (C.P.); lixifeng@shu.edu.cn (X.L.)

Abstract: In this paper, the effect of a buffer layer created using different hydrogen-containing ratios of reactive gas on the electrical properties of a top-gate In-Ga-Zn-O thin-film transistor was thoroughly investigated. The interface roughness between the buffer layer and active layer was characterized using atomic force microscopy and X-ray reflection. The results obtained using Fourier transform infrared spectroscopy show that the hydrogen content of the buffer layer increases with the increase in the hydrogen content of the reaction gas. With the increase in the hydrogen-containing materials in the reactive gas, field effect mobility and negative bias illumination stress stability improve nearly twofold. The reasons for these results are explained using technical computer-aided design simulations.

Keywords: top-gate IGZO TFT; hydrogen-containing; buffer layer; technical computer-aided design simulation

check for updates

Citation: Huang, H.; Peng, C.; Xu, M.; Chen, L.; Li, X. Dependence of a Hydrogen Buffer Layer on the Properties of Top-Gate IGZO TFT. *Micromachines* **2024**, *15*, 722. https://doi.org/10.3390/mi15060722

Academic Editor: Chengyuan Dong

Received: 14 May 2024
Revised: 28 May 2024
Accepted: 28 May 2024
Published: 29 May 2024

Copyright: © 2024 by the authors. Licensee MDPI, Basel, Switzerland. This article is an open access article distributed under the terms and conditions of the Creative Commons Attribution (CC BY) license (https://creativecommons.org/licenses/by/4.0/).

1. Introduction

In recent investigations of advanced semiconductors, amorphous oxide semiconductors, especially In-Ga-Zn-O (IGZO) ones, have been widely studied in relation to their ability to act as active layer materials for thin-film transistors (TFTs) due to their high field-effect mobility (μ), large-area uniformity (>Generation 8; 2200 mm × 2500 mm), low leakage current, low-temperature (<300 °C) fabrication process, and excellent transparency in the visible region [1–6]. At present, the structure of IGZO TFT can be roughly divided into a bottom gate and a top gate based on the active layer [3,5–7]. For the bottom-gate structure IGZO TFT, highly energetic particles generated in the deposition process used for a semiconductor channel (such as sputtering) are likely to cause damage to the dielectric. However, the dielectric in a top-gate IGZO TFT can serve as a gas permeation barrier [8]. In addition, a top-gate IGZO TFT is considered to be the most suitable structure for large high-resolution panel displays because it can provide better process controllability [7,9,10]. So, top-gate IGZO TFTs are receiving more and more attention from industry and academia [5,7,9–11]. In addition, IGZO films have become the most promising semiconductor materials in the flexible display field. Aluminum, stainless steel, and polyimide/polyethylene naphthalate have some issues when serving as the substrates of flexible displays, such as the electrical conductivity of aluminum and stainless steel and polyimide/polyethylene naphthalate's poor adhesion to the device layer [12–14]. At the same time, there may be a stress mismatch between the substrate and the device layer. Therefore, to improve the applicability of top-gate IGZO TFTs, it is usually necessary to deposit a buffer layer on the substrate before fabricating TFT devices [12,13]. Typical buffer layers are composed of silicon dioxide (SiO_2) or silicon nitride (Si_3N_4) deposited via plasma chemical vapor deposition (PECVD), and their preparation generally requires using some special gases as reaction sources, such as NH_3, SiH_4, and so on [3,7].

During the growth of the buffer layer, a large amount of hydrogen will be introduced into the film, and it is difficult to precisely control the hydrogen content within a reasonable range [7]. Some previous studies have shown that the hydrogen content of the buffer layer can be changed by changing how the buffer layers are stacked, the annealing temperature, or the type of substrate [2,13–15]. In addition, they have also put forward the idea that hydrogen can act as an impurity with a shallow donor state, and an appropriate hydrogen concentration in the buffer layer can improve the performance of devices such that they meet the requirements of display driving [7,9,13–18]. However, few studies have explored the preparation process for the precise control of the hydrogen content of the buffer layer, which affects the performance of a device. At the same time, there are hardly any reports that explain the effect of a buffer layer created using different hydrogen-containing ratios of reactive gas on the electrical properties of a top-gate IGZO TFT according to the density of states (DOSs) extracted using computer-aided design (TCAD) simulation.

In this study, we adopted the method of adjusting the hydrogen-containing ratio of the reactive gas for the buffer layer to precisely control the performance of top-gate IGZO TFTs. The corresponding relationship between the hydrogen-containing ratio of the reactive gas and the hydrogen content in the buffer layer was ascertained via nondestructive Fourier transform infrared spectroscopy. The DOS of the channel layer was deduced based on TCAD simulation.

2. Experiment

Staggered top-gate bottom-contact TFTs with IGZO channel layers were constructed on glass substrates. A schematic cross-sectional diagram and optical top view of the device are shown in Figure 1a,b, respectively. First, a 200 nm thick buffer layer was deposited on a 200 × 200 mm sheet of glass via PECVD, including SiO_2 and Si_3N_4 in the process. Following this, a sheet of indium tin oxide (ITO) with a thickness of 35 nm and a sheet of IGZO with a thickness of 40 nm were sputtered via magnetically controlled sputtering as source/drain electrode (S/D) and active layers, respectively. Then, a SiO_2 sheet with a thickness of 300 nm was successively deposited as a gate-insulating layer (GI) using PECVD. Finally, a 35 nm thick ITO thin film was sputtered again as a gate electrode (G) using magnetron sputtering. In addition, the patterning of each layer was achieved using a conventional lithography process. The width/length ratio (W/L) of all devices was 8/8 μm μm^{-1}.

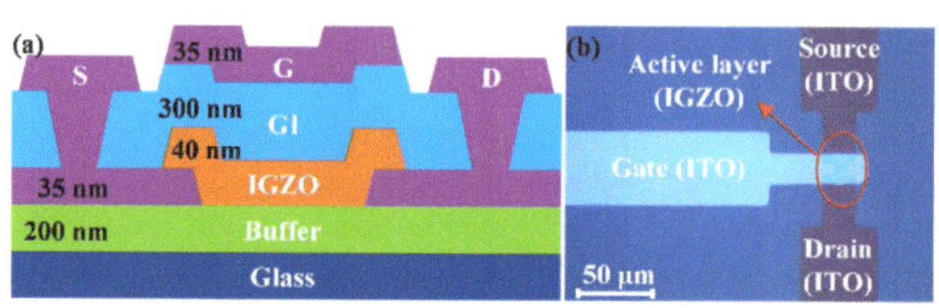

Figure 1. (**a**) The schematic diagram of top-gate IGZO TFTs. (**b**) Optical top view of top-gate IGZO TFT.

The surface morphology of the film was analyzed using an atomic force microscope (AFM, Bruker, Karlsruhe, Germany). The buffer/active layer interface roughness was analyzed using *X*-ray reflectivity (XRR, Smart lab, Tokyo, Japan). The hydrogen content of the buffer films was analyzed and calculated using Fourier transform infrared (FTIR, Nicolet 380, Thermo Fisher Scientific, Waltham, MA, USA) spectroscopy. The electrical performance of the devices was tested using a Keithley 4200 semiconductor, Tektronix, Beaverton, OR, USA) parameter analyzer. The transfer characteristics of all transfers were measured at a drain voltage of 10 V. The gate bias tests used were the negative bias stress (NBS) and negative bias illumination stress (NBIS) tests. The light source was a light-emitting diode (LED), whose light intensity was 10,000 lux, and the corresponding spectrum is shown in Figure 2. The threshold voltage (V_{TH}) was determined from the x-axis intercept of the $I_{DS}^{1/2}$ versus V_{GS} plot using the linear extrapolation method. The μ was calculated according to the following equation:

$$\mu = 2L \cdot I_{DS} / W \cdot C_i \cdot (V_{GS} - V_{TH})^2$$

C_i is the gate capacitance per unit area.

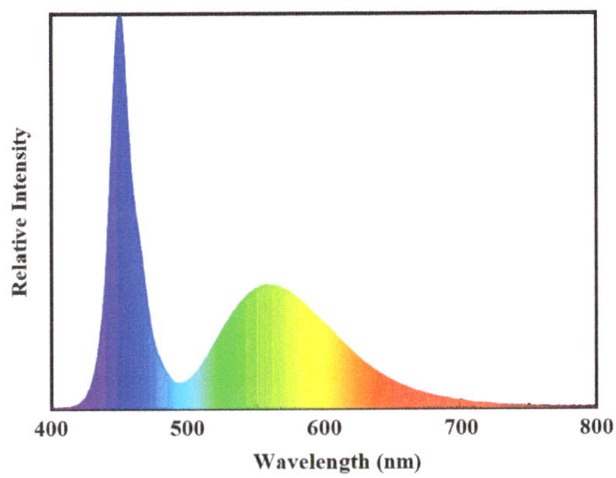

Figure 2. The emission spectra of the white LED backlight.

3. Results and Discussion

3.1. Thin-Film Performance Analysis

To investigate the influence of the different hydrogen-containing ratios of the reactive gas in the buffer layer on the performance of thin films, the buffer layer films with a thickness of 200 nm were deposited on double-sided polished silicon wafers using PECVD (ULVAC, CME-200E) at 200 °C. As shown in Table 1, according to the hydrogen-containing ratios of the reactive gases in the developed buffer layer, the buffer layer is represented by normalized N_{H0}, N_{H3}, N_{H28}, N_{H93}, and N_{H100}, respectively.

Table 1. Values that were normalized to represent the hydrogen-containing ratios of reactive gas for the buffer layer.

Buffer	Reactive Gas	Reaction Gas Ratio	H%
w/o	Without	Without	N_{H0}
SiO_2	SiH_4/N_2O	4/700 44/660 144/560	N_{H3} N_{H28} N_{H93}
Si_3N_4	$SiH_4/NH_3/N_2$	40/154/510	N_{H100}

Figure 3a–d show the variation in the surface morphology of the buffer layer films with the hydrogen content of the growth gas obtained via atomic force microscopy (AFM). It can be seen that the surface root-mean-square roughness (RMS) increases from 0.16 nm to 0.25 nm as the hydrogen content increases from N_{H3} to N_{H100}. The hydrogen concentration increased, which implies that some of the Si-O-Si bonding was replaced by H-terminated Si-OH and Si-H bonding on the surface of the buffer layer, enhancing the chemical reactivity of the silica surface and making it easier to oxidize or reduce the surface [19–21]. In addition, the increase in surface Si-H bonds makes it easier to form a loose porous structure, which affects the structure and morphology of the buffer layer thin-film surface, increasing surface roughness [21–24].

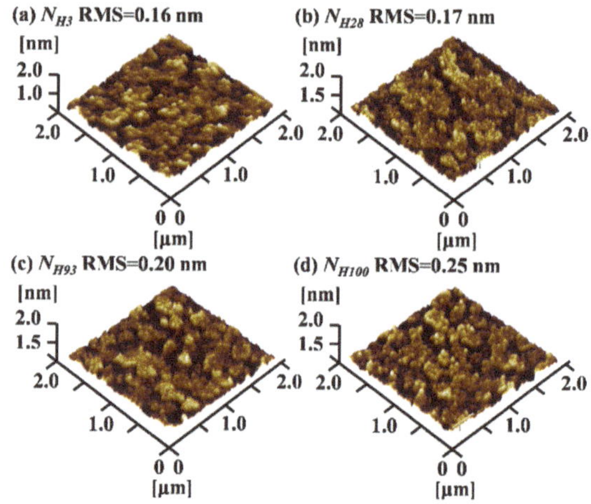

Figure 3. AFM 3D images of the morphology of the buffer layers: (**a**) N_{H3}, (**b**) N_{H28}, (**c**) N_{H93}, and (**d**) N_{H100}.

In particular, N_{H3} and N_{H100} buffer layer films were chosen as representative subjects based on the hydrogen-containing ratio of reactive gas. A 40 nm thick IGZO film was deposited on top of them (N_{H3}/IGZO and N_{H100}/IGZO), and then XRR was used to preliminarily determine the hydrogen-containing ratio of the reactive gas in the buffer layer on the interface between the buffer layer and the active layer. Figure 4 shows the measured (black solid line) and simulated (red solid line) XRR results for (a) N_{H3}/IGZO and (b) N_{H100}/IGZO. It can be seen that the simulation curve of the N_{H3}/IGZO sample is smoother than that of N_{H100}/IGZO, indicating that the interface of the former is smoother, and it produces less surface carrier scattering [25], which agrees sufficiently well with the AFM results. The IGZO densities of the N_{H3}/IGZO and N_{H100}/IGZO samples were 6.17 g/cm^{-3} and 6.15 g/cm^{-3}, respectively, and this extremely small error was caused by system measurement or calculation errors, so the change in hydrogen content in the buffer layer did not cause the density change of the IGZO layer.

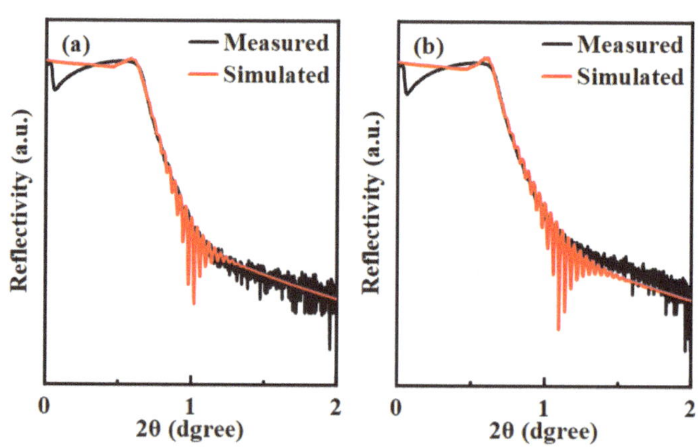

Figure 4. Measured (black solid line) and simulated (red solid lines) XRR spectra for the IGZO layers deposited on different buffer layers: (**a**) N_{H3}/IGZO; (**b**) N_{H100}/IGZO.

To clarify the corresponding relationship between the hydrogen-containing ratio of reactive gases for the buffer layer and the hydrogen content of the buffer layer film, the films deposited with different hydrogen-containing ratios of reactive gas were analyzed by FTIR. The absorption intensity of the FTIR absorption spectrum is positively correlated with the hydrogen content in the film, and the absorption peak near 640 cm^{-1} includes the wagging-rocking modes of Si-H wagging vibration [26]. Therefore, the hydrogen content is generally expressed by the intensity of the corresponding peak at 640 cm^{-1} [26,27]. The C_H is determined using the following relationship: $C_H = A_\omega \cdot I(\omega)/N$, $I(\omega) = \int [\alpha(\omega)/\omega]d\omega$; here, A_{640} is the proportionality constant for this Si-H mode, with the value used for the films studied in this work being 1.6×10^{19} cm^{-2}, and N is the atomic density of silicon atoms in c-Si, which is taken to be 5.0×10^{22} cm^{-3} [27]. Figure 5 depicts the absorption spectrum and Gaussian fitting results of the FTIR spectrum around 640 cm^{-1}. As the ratio of hydrogen-containing reactive gases in the grown buffer layer increases from N_{H0} to N_{H100}, the hydrogen content in the buffer layer film increases from 4.04 at% to 21.60 at%, as the higher the hydrogen content in the reaction gas, the higher the hydrogen content in the film. However, after annealing the buffer layer film of N_{H100}, the hydrogen content decreased from 21.60 at% to 2.94 at%, which can be attributed to the diffusion of highly active hydrogen away from the buffer layer film [2,14,28,29].

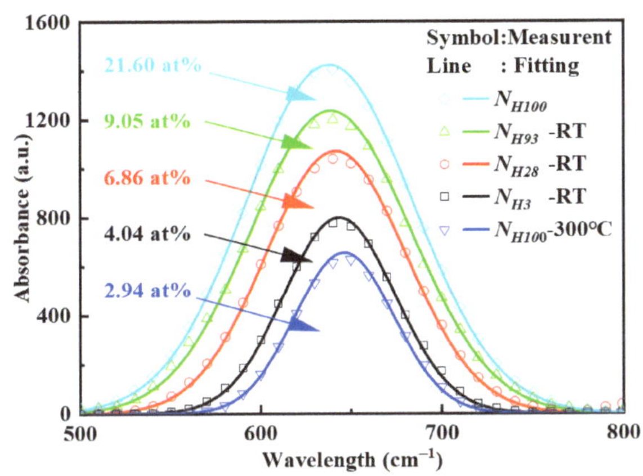

Figure 5. FTIR spectra at 640 cm^{-1} of the film deposited at different hydrogen−containing gas ratios of the buffer layers.

3.2. The Influence of Hydrogen Content in the Buffer Layer

Figure 6 shows the transfer characteristic curves of the top-gate IGZO TFT corresponding to different buffer hydrogen proportions after annealing. The device performance values are summarized in Table 2. As the ratio of hydrogen-containing reactive gases in the growth buffer layer increases from N_{H0} to N_{H100}, the buffer capacitance per unit area (C_i) increases from 16.83 to 31.45 nF/cm^2, μ monotonically increases from 4.29 cm^{2}V^{-1}s^{-1} to 11.46 cm^2/V·s, the on/off ratio for current (I_{on}/I_{off}) slowly increases from 1.14×10^8 to 2.33×10^9, the subthreshold swing (SS) increases from 0.16 V/dec to 0.86 V/dec, and V_{TH} gradually shifts leftward from 7.78 V to −0.73 V.

Table 2. Electrical characteristics of top-gate IGZO TFTs with different hydrogen-containing gas ratios in the buffer layers.

Buffer	H%	C_i (nF/cm^2)	μ (cm^2/V·s)	I_{on}/I_{off}	SS (V/dec)	V_{TH} (V)
w/o	N_{H0}	16.83 ± 0.03	4.29 ± 0.34	$1.14 \times 10^8 \pm 4.23 \times 10^7$	0.16 ± 0.04	7.78 ± 0.33
SiO_2	N_{H3}	17.28 ± 0.04	5.74 ± 0.28	$2.67 \times 10^8 \pm 1.57 \times 10^7$	0.21 ± 0.02	4.79 ± 0.26
	N_{H28}	17.50 ± 0.04	7.85 ± 0.23	$6.14 \times 10^8 \pm 3.74 \times 10^7$	0.23 ± 0.02	0.76 ± 0.30
	N_{H93}	18.16 ± 0.02	8.24 ± 0.17	$9.21 \times 10^8 \pm 5.87 \times 10^7$	0.34 ± 0.03	−0.56 ± 0.15
Si_3N_4	N_{H100}	31.45 ± 0.05	11.46 ± 0.15	$2.33 \times 10^9 \pm 3.66 \times 10^8$	0.86 ± 0.07	−0.73 ± 0.21

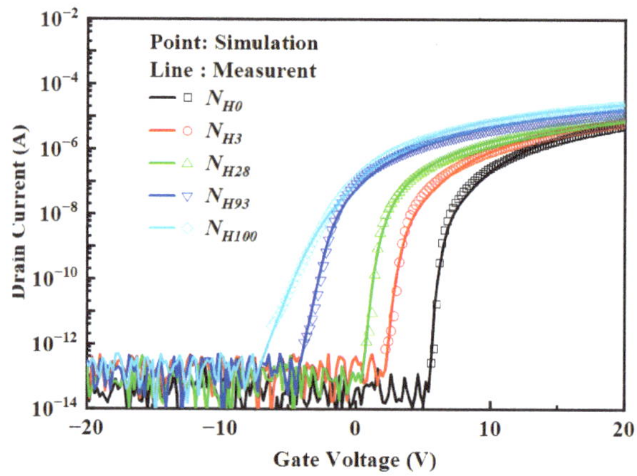

Figure 6. Comparison of experimental and simulated transfer characteristics.

Since the hydrogen in the adjacent layers of IGZO will diffuse to the active layer, the hydrogen diffusion model shown in Figure 7a was established. The active layer was simplified to two equivalent resistances, the active layer resistance affected by the insulating layer is defined as $R_{CH\text{-}Top}$, and the active layer resistance affected by the buffer layer is defined as $R_{CH\text{-}Bottom}$, as shown in Figure 7b. Because the insulating layer growth process is the same for all top-gate IGZO TFTs, the hydrogen content diffused from the insulating layer into the IGZO can be considered to be approximately the same, and therefore the $R_{CH\text{-}Top}$ is the same for all the devices. Therefore, only the influence of hydrogen in the buffer layer on the performance of IGZO was considered. The hydrogen atoms in the buffer layer will diffuse into IGZO through the Buffer/IGZO interface and then combine with O^{2-} ions in the IGZO film to form hydroxyl groups, which are released electrons. As the hydrogen-containing ratio of reactive gases increases, the number of hydrogen atoms diffused into the IGZO also increases, so the electron concentration of the lower IGZO layer increases and the $R_{CH\text{-}Bottom}$ decreases, which is beneficial to the conduction of electrons. In general, the resistance between the source and drain electrodes is generated by the parallel resistance of the upper and lower layers of IGZO, so as the hydrogen content of the buffer layer increases, the total resistance between the source and drain electrodes will become smaller [30]. At the same time, the increase in the electron concentration of the entire channel will cause the device to be turned on in advance so that the V_{TH} drifts to the left [13]. In addition, a higher carrier concentration also increases the mobility of IGZO [1]. With the increase in H concentration, donor and acceptor effects alternately play a leading role, as shown by the formation of the H_2 molecule (-OH-H), which leads to the fluctuation of electrical parameters. In addition, excessive H can replace O in weak metal-oxygen bonds, which inhibits the bonding of interface metal-oxygen bonds and increases the number of interface defects, leading to the deterioration of *SS* [31,32].

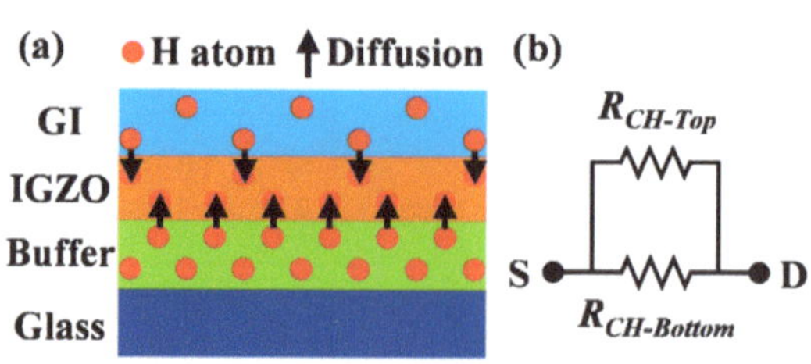

Figure 7. (**a**) Schematic diagram of hydrogen diffusion in adjacent layers of active layer. (**b**) Equivalent resistance model.

3.3. 2D Numerical Simulation

A Silvaco ATLAS 2-D device simulator was used to investigate the differences in top-gate IGZO TFT devices with the different hydrogen-containing ratios of reactive gases for the buffer layer, particularly the difference in the DOS. Figure 6 shows the experimental and simulated transfer characteristics. Excellent agreement between the experiment and simulation was achieved. The sub-bandgap state nomenclature employed in Table 3 can be explained as follows. The acceptor-like tail states are defined by the peak density N_{TA} and the Urbach energy (slope) W_{TA}. The donor-like deep-level defect states are defined by the peak density N_{GD}, the characteristic decay energy W_{GD}, and the peak energy E_{GD}. The density of fixed charges is represented by QF. The DOS and the key parameters of the defect model have the following relationship [33]:

$$\text{DOS} = N_{TA}\cdot\exp[(E - E_C)/W_{TA}] + N_{GD}\cdot\exp[-(E - E_{GD})^2/W_{GD}{}^2]$$

Table 3. Densities of key defect model parameters for top-gate IGZO TFTs fitted according to different hydrogen-containing gas ratios of the buffer layers.

H%	N_{TA} $cm^{-3}eV^{-1}$	W_{TA} eV	N_{GD} $cm^{-3}eV^{-1}$	W_{GD} eV	E_{GD} eV	QF cm^{-2}
N_{H0}	1.57×10^{20}	0.032	3.00×10^{17}	0.12	2.72	2.30×10^{11}
N_{H3}	1.55×10^{20}	0.032	3.20×10^{17}	0.12	2.70	2.60×10^{11}
N_{H28}	1.40×10^{20}	0.032	3.50×10^{17}	0.12	2.68	3.00×10^{11}
N_{H93}	1.00×10^{20}	0.032	4.00×10^{17}	0.12	2.67	4.50×10^{11}
N_{H100}	6.00×10^{19}	0.032	4.30×10^{17}	0.12	2.65	5.00×10^{11}

Device performance was controlled by adjusting the hydrogen content in the buffer layer and controlling the diffusion of hydrogen-related impurities to adjust the hydrogen content in the IGZO. As the hydrogen-containing ratio of reactive gases increases from N_{H0} to N_{H100}, I_{on} increases from 5.79×10^{-6} A to 2.46×10^{-5} A, which can be attributed to the decrease in N_{TA} from 1.57×10^{20} $cm^{-3}eV^{-1}$ to 6.00×10^{19} $cm^{-3}eV^{-1}$. The decrease in N_{TA} means a decrease in acceptor-like tail states, which mainly capture free electrons transitioning to the conduction band so that there will be more free electrons transitioning to the conduction band at the same gate voltage [18,34]. In addition, the free electrons in the conduction band are conducted in the extended state, while the electron conduction in the band tail state consists of hopping conduction limited by traps, and its conductivity is much smaller than that of the extended state, so the electrons in the band tail state are conducted via hopping conduction limited by traps. The decrease in the concentration of trapped electrons will inevitably increase the effective mobility of electrons and increase I_{on} [35,36]. While the hydrogen-containing ratio of reactive gas for the buffer layer increases from N_{H0} to N_{H100}, the SS increases from 0.16 V/dec to 0.86 V/dec, which can be attributed to the increase in N_{GD} from 3.00×10^{17} $cm^{-3}eV^{-1}$ to 4.30×10^{17} $cm^{-3}eV^{-1}$, and QF increases from 2.30×10^{11} cm^{-2} to 5.00×10^{11} cm^{-2}. The change in SS is not substantially related to the change of acceptor-like tail states, mainly because acceptor-like tail states are closer to the conduction band, and their change does not affect the subthreshold region.

As the hydrogen-containing ratio of reactive gases increases from N_{H0} to N_{H100}, the V_{TH} shifts continuously to the left from 7.78 V to −0.73 V, which may be due to the increase in N_{GD} from 3.00×10^{17} $cm^{-3}eV^{-1}$ to 4.30×10^{17} $cm^{-3}eV^{-1}$. An increase in N_{GD} implies an increase in donor-like deep-level defects [18,37,38]. At the same voltage, the number of free electrons generated increases, and the number of electrons that can transfer to the conduction band also increases. This enables the device to turn on at a more negative gate voltage, so V_{TH} shifts to the left [35,38,39]. The E_{GD} decreased from 2.72 eV to 2.65 eV, indicating that the oxygen vacancy defect energy level shifted to the valence band, and the distance from the defect energy level to the bottom of the conduction band increased, which improved the NBIS stability of the device [20,40].

3.4. The Influence of Hydrogen Content on Stability

Figure 8 depicts the stability of top-gate IGZO TFTs with different hydrogen-containing ratios of reactive gases for the buffer layer under NBIS. With the increase in the hydrogen-containing ratio of reactive gases, ΔV_{TH} reduced from −3.27 V to −1.21 V. Due to the presence of electrically neutral donor-like defect states introduced by oxygen-related defects in IGZO, the donor-like defect states will release electrons under NBIS and increase the carrier concentration in the channel, so the threshold voltage under NBIS will shift to the left. On the one hand, the V_O increases through the combination of hydrogen atoms with O^{2-} [41]. On the other hand, the hydrogen atom forms a substitutional impurity, forming a stable metal–hydrogen bond with metal ions, and the distance from the defect level to the bottom of the conduction band also increases, so the NBIS of the top-gate IGZO TFT improves [4,42,43].

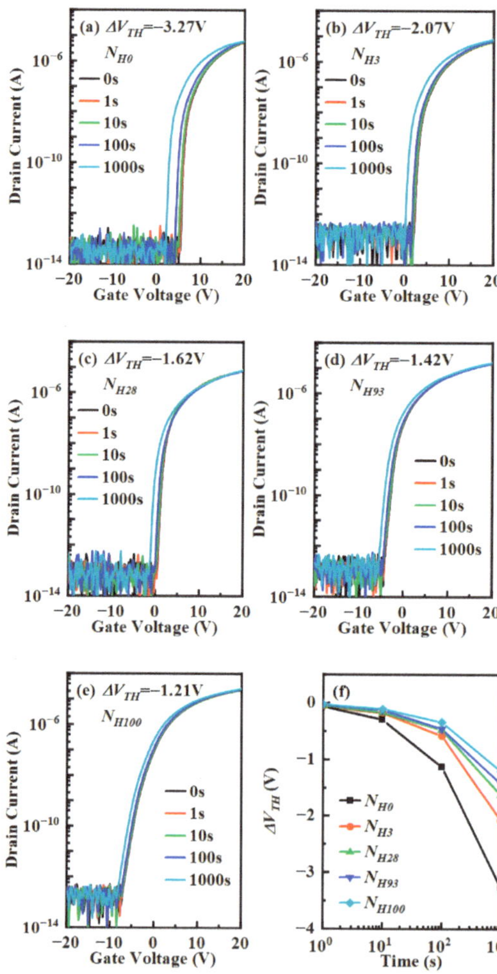

Figure 8. NBIS stability for top-gate IGZO TFT under different buffer hydrogen proportions: (**a**) N_{H0}, (**b**) N_{H3}, (**c**) N_{H28}, (**d**) N_{H93}, (**e**) N_{H100}, and (**f**) ΔV_{TH} as a function of stress time.

4. Conclusions

In this paper, top-gate IGZO TFTs with different hydrogen proportions in the buffer layer were successfully fabricated, and the effect of hydrogen content on the stability of negative bias illumination stress was discussed. It has been found that the results of atomic force microscopy and *X*-ray reflection indicate that Si_3N_4 films with higher hydrogen content have greater surface and Si_3N_4/IGZO interface roughness, respectively. By optimizing the hydrogen content of the buffer layer, the field-effect mobility improved nearly threefold, reaching 11.46 $cm^2/V{\cdot}s$, while the NBIS stability was remarkably enhanced. TACD simulations further confirmed that deep donor-like and acceptor-like defects can be controlled by the hydrogen-containing ratio of reactive gases, consulting the reason for the remarkable performance of top-gate IGZO TFTs.

Author Contributions: Conceptualization, X.L. and C.P.; methodology, C.P. and X.L.; funding acquisition, X.L. and C.P.; investigation, C.P., H.H. and M.X.; writing—original draft preparation, H.H. and C.P.; writing—review and editing, C.P., L.C. and X.L.; supervision, X.L. All authors have read and agreed to the published version of the manuscript.

Funding: This work is supported by the National Natural Science Foundation of China under Grants U22A6002, 62174105 and 62304128.

Data Availability Statement: The data and contributions presented in the study are included in this article. Further inquiries can be directed to the corresponding author.

Conflicts of Interest: The authors declare no conflicts of interest.

References

1. Park, J.C.; Ahn, S.-E.; Lee, H.-N. High-Performance Low-Cost Back-Channel-Etch Amorphous Gallium–Indium–Zinc Oxide Thin-Film Transistors by Curing and Passivation of the Damaged Back Channel. *ACS Appl. Mater. Interfaces* **2013**, *5*, 12262–12267. [CrossRef] [PubMed]
2. Han, K.-L.; Ok, K.-C.; Cho, H.-S.; Oh, S.; Park, J.-S. Effect of Hydrogen on the Device Performance and Stability Characteristics of Amorphous InGaZnO Thin-Film Transistors with a $SiO_2/SiN_x/SiO_2$ Buffer. *Appl. Phys. Lett.* **2017**, *111*, 063502. [CrossRef]
3. Peng, C.; Yang, S.; Pan, C.; Li, X.; Zhang, J. Effect of Two-Step Annealing on High Stability of a-IGZO Thin-Film Transistor. *IEEE Trans. Electron. Devices* **2020**, *67*, 4262–4268. [CrossRef]
4. Zhang, Y.; He, G.; Wang, L.; Wang, W.; Xu, X.; Liu, W. Ultraviolet-Assisted Low-Thermal-Budget-Driven α-InGaZnO Thin Films for High-Performance Transistors and Logic Circuits. *ACS Nano* **2022**, *16*, 4961–4971. [CrossRef] [PubMed]
5. Peng, C.; Xu, M.; Chen, L.; Li, X.; Zhang, J. Improvement of Properties of Top-Gate IGZO TFT by Oxygen-Rich Ultrathin in Situ ITO Active Layer. *Jpn. J. Appl. Phys.* **2022**, *61*, 070914. [CrossRef]
6. Lin, D.; Su, W.-C.; Chang, T.-C.; Chen, H.-C.; Tu, Y.-F.; Zhou, K.-J.; Hung, Y.-H.; Yang, J.; Lu, I.-N.; Tsai, T.-M. Degradation Behavior of Etch-Stopper-Layer Structured a-InGaZnO Thin-Film Transistors under Hot-Carrier Stress and Illumination. *IEEE Trans. Electron. Devices* **2021**, *68*, 556–559. [CrossRef]
7. Song, A.; Hong, H.M.; Son, K.S.; Lim, J.H.; Chung, K.-B. Hydrogen Behavior in Top Gate Amorphous In-Ga-Zn-O Device Fabrication Process during Gate Insulator Deposition and Gate Insulator Etching. *IEEE Trans. Electron. Devices* **2021**, *68*, 2723–2728. [CrossRef]
8. Fakhri, M.; Theisen, M.; Behrendt, A.; Görrn, P.; Riedl, T. Top-Gate Zinc Tin Oxide Thin-Film Transistors with High Bias and Environmental Stress Stability. *Appl. Phys. Lett.* **2014**, *104*, 251603. [CrossRef]
9. Lee, M.-X.; Chiu, J.-C.; Li, S.-L.; Sarkar, E.; Chen, Y.-C.; Yen, C.-C.; Chen, T.-L.; Chou, C.-H.; Liu, C. Mobility Enhancement and Abnormal Humps in Top-Gate Self-Aligned Double-Layer Amorphous InGaZnO TFTs. *IEEE J. Electron. Devices Soc.* **2022**, *10*, 301–308. [CrossRef]
10. Hong, S.-Y.; Kim, H.-J.; Kim, D.-H.; Jeong, H.-Y.; Song, S.-H.; Cho, I.-T.; Noh, J.; Yun, P.S.; Lee, S.-W.; Park, K.-S. Study on the Lateral Carrier Diffusion and Source-Drain Series Resistance in Self-Aligned Top-Gate Coplanar InGaZnO Thin-Film Transistors. *Sci. Rep.* **2019**, *9*, 6588. [CrossRef]
11. Chen, H.-C.; Chen, G.-F.; Chen, P.-H.; Huang, S.-P.; Chen, J.-J.; Zhou, K.-J.; Kuo, C.-W.; Tsao, Y.-C.; Chu, A.-K.; Huang, H.-C. A Novel Heat Dissipation Structure for Inhibiting Hydrogen Diffusion in Top-Gate a-InGaZnO TFTs. *IEEE Electron. Device Lett.* **2019**, *40*, 1447–1450. [CrossRef]
12. Petti, L.; Münzenrieder, N.; Vogt, C.; Faber, H.; Büthe, L.; Cantarella, G.; Bottacchi, F.; Anthopoulos, T.D.; Tröster, G. Metal Oxide Semiconductor Thin-Film Transistors for Flexible Electronics. *Appl. Phys. Rev.* **2016**, *3*, 021303. [CrossRef]
13. Han, K.-L.; Han, J.-H.; Kim, B.-S.; Jeong, H.-J.; Choi, J.-M.; Hwang, J.-E.; Oh, S.; Park, J.-S. Organic/Inorganic Hybrid Buffer in InGaZnO Transistors under Repetitive Bending Stress for High Electrical and Mechanical Stability. *ACS Appl. Mater. Interfaces* **2019**, *12*, 3784–3791. [CrossRef] [PubMed]

14. Han, K.-L.; Cho, H.-S.; Ok, K.-C.; Oh, S.; Park, J.-S. Comparative Study on Hydrogen Behavior in InGaZnO Thin Film Transistors with a $SiO_2/SiN_x/SiO_2$ Buffer on Polyimide and Glass Substrates. *Electron. Mater. Lett.* **2018**, *14*, 749–754. [CrossRef]
15. Ok, K.-C.; Ko Park, S.-H.; Hwang, C.-S.; Kim, H.; Soo Shin, H.; Bae, J.; Park, J.-S. The Effects of Buffer Layers on the Performance and Stability of Flexible InGaZnO Thin Film Transistors on Polyimide Substrates. *Appl. Phys. Lett.* **2014**, *104*, 063508. [CrossRef]
16. Chen, C.; Yang, B.; Li, G.; Zhou, H.; Huang, B.; Wu, Q.; Zhan, R.; Noh, Y.; Minari, T.; Zhang, S. Analysis of Ultrahigh Apparent Mobility in Oxide Field-effect Transistors. *Adv. Sci.* **2019**, *6*, 1801189. [CrossRef]
17. Nakashima, M.; Oota, M.; Ishihara, N.; Nonaka, Y.; Hirohashi, T.; Takahashi, M.; Yamazaki, S.; Obonai, T.; Hosaka, Y.; Koezuka, J. Origin of Major Donor States in In-Ga-Zn Oxide. *J. Appl. Phys.* **2014**, *116*, 213703. [CrossRef]
18. Noh, H.Y.; Kim, J.; Kim, J.-S.; Lee, M.-J.; Lee, H.-J. Role of Hydrogen in Active Layer of Oxide-Semiconductor-Based Thin Film Transistors. *Crystals* **2019**, *9*, 75. [CrossRef]
19. Park, H.; Yun, J.; Park, S.; Ahn, I.; Shin, G.; Seong, S.; Song, H.-J.; Chung, Y. Enhancing the Contact between A-IGZO and Metal by Hydrogen Plasma Treatment for a High-Speed Varactor (>30 GHz). *ACS Appl. Electron. Mater.* **2022**, *4*, 1769–1775. [CrossRef]
20. Bang, J.; Matsuishi, S.; Hosono, H. Hydrogen Anion and Subgap States in Amorphous In-Ga-Zn-O Thin Films for TFT Applications. *Appl. Phys. Lett.* **2017**, *110*, 232105. [CrossRef]
21. Su, W.-S.; Fang, W.; Tsai, M.-S. Tuning the Mechanical Properties of SiO_2 Thin Film for MEMS Application. *MRS Online Proc. Lib.* **2003**, *795*, 487–492. [CrossRef]
22. Ashby, M.F. Overview No. 80: On the Engineering Properties of Materials. *Acta Metall.* **1989**, *37*, 1273–1293. [CrossRef]
23. Park, S.; Park, T.; Choi, Y.; Jung, C.; Kim, B.; Jeon, H. Radical-Induced Effect on PEALD SiO_2 Films by Applying Positive DC Bias. *ECS J. Solid State Sci. Technol.* **2022**, *11*, 023007. [CrossRef]
24. Lin, D.; Yang, J.-Z.; Cheng, J.-R.; Deng, X.-C.; Chen, Y.-S.; Zhuang, P.-P.; Li, T.-J.; Liu, J. InSnO: N Homojunction Thin-Film Transistors Fabricated at Room Temperature. *Vacuum* **2023**, *213*, 112099. [CrossRef]
25. Jeong, H.; Nam, S.; Park, K.; Choi, H.; Jang, J. Finding the Cause of Degradation of Low-Temperature Oxide Thin-Film Transistors. *J. Korean Phys. Soc* **2021**, *78*, 284–289. [CrossRef]
26. Wang, S.-H.; Chang, H.-E.; Lee, C.-C.; Fuh, Y.-K.; Li, T.T. Evolution of A-Si:H to Nc-Si:H Transition of Hydrogenated Silicon Films Deposited by Trichlorosilane Using Principle Component Analysis of Optical Emission Spectroscopy. *Mater. Chem. Phys.* **2020**, *240*, 122186. [CrossRef]
27. Goh, B.T.; Wah, C.K.; Aspanut, Z.; Rahman, S.A. Structural and Optical Properties of Nc-Si: H Thin Films Deposited by Layer-by-Layer Technique. *J. Mater. Sci. Mater. Electron.* **2014**, *25*, 286–296. [CrossRef]
28. Aman, S.M.; Koretomo, D.; Magari, Y.; Furuta, M. Influence of Deposition Temperature and Source Gas in PE-CVD for SiO_2 Passivation on Performance and Reliability of In-Ga-Zn-O Thin-Film Transistors. *IEEE Trans. Electron. Devices* **2018**, *65*, 3257–3263. [CrossRef]
29. Wu, Y.; Lan, L.; He, P.; Lin, Y.; Deng, C.; Chen, S.; Peng, J. Influence of Hydrogen Ions on the Performance of Thin-Film Transistors with Solution-Processed AlOx Gate Dielectrics. *Appl. Sci.* **2021**, *11*, 4393. [CrossRef]
30. Zhou, L.; Guo, X.; Ouyang, B.; Wang, M.; Ma, Q.; Wang, B. Amorphous IGZO Thin-Film Transistor Gate Driver in Array for Ultra-Narrow Border Displays. *IEEE J. Electron Devices Soc.* **2022**, *10*, 351–355. [CrossRef]
31. Chowdhury, M.D.H.; Mativenga, M.; Um, J.G.; Mruthyunjaya, R.K.; Heiler, G.N.; Tredwell, T.J.; Jang, J. Effect of SiO_2 and SiO_2/SiN_x Passivation on the Stability of Amorphous Indium-Gallium Zinc-Oxide Thin-Film Transistors under High Humidity. *IEEE Trans. Electron. Devices* **2015**, *62*, 869–874. [CrossRef]
32. Pereira, M.E.; Deuermeier, J.; Freitas, P.; Barquinha, P.; Zhang, W.; Martins, R.; Fortunato, E.; Kiazadeh, A. Tailoring the Synaptic Properties of A-IGZO Memristors for Artificial Deep Neural Networks. *APL Mater.* **2022**, *10*, 011113. [CrossRef]
33. Zhang, P.; Samanta, S.; Fong, X. Physical Insights into the Mobility Enhancement in Amorphous InGaZnO Thin-Film Transistor by SiO_2 Passivation Layer. *IEEE Trans. Electron. Devices* **2020**, *67*, 2352–2358. [CrossRef]
34. Zhu, Z.; Cao, W.; Huang, X.; Shi, Z.; Zhou, D.; Xu, W. Analysis of Nitrogen-Doping Effect on Sub-Gap Density of States in a-IGZO TFTs by TCAD Simulation. *Micromachines* **2022**, *13*, 617. [CrossRef] [PubMed]
35. Soufyane, N.; Sengouga, N.; Labed, M.; Meftah, A. Temperature Dependent Poly Crystalline Zinc Oxide Thin Film Transistor Characteristics. *Trans. Electr. Electron. Mater.* **2021**, *22*, 711–716. [CrossRef]
36. Kumar, N.; Sutradhar, M.; Kumar, J.; Panda, S. Role of Deposition and Annealing of the Top Gate Dielectric in A-IGZO TFT-Based Dual-Gate Ion-Sensitive Field-Effect Transistors. *Semicond. Sci. Technol.* **2017**, *32*, 035013. [CrossRef]
37. Peng, H.; Chang, B.; Fu, H.; Yang, H.; Zhang, Y.; Zhou, X.; Lu, L.; Zhang, S. Top-Gate Amorphous Indium-Gallium-Zinc-OxideThin-Film Transistors With Magnesium Metallized Source/Drain Regions. *IEEE Trans. Electron. Devices* **2020**, *67*, 1619–1624. [CrossRef]
38. Billah, M.M.; Chowdhury, M.D.H.; Mativenga, M.; Um, J.G.; Mruthyunjaya, R.K.; Heiler, G.N.; Tredwell, T.J.; Jang, J. Analysis of Improved Performance under Negative Bias Illumination Stress of Dual Gate Driving A-IGZO TFT by TCAD Simulation. *IEEE Electron. Device Lett.* **2016**, *37*, 735–738. [CrossRef]
39. Raj, R.B.; Tripathi, A.K.; Mahato, P.K.; Nair, S.; Shahana, T.; Mukundan, T. Effect of Active Layer Thickness Variation on Scaling Response in A-IGZO Thin Film Transistors under Schottky Limited Operation. *Semicond. Sci. Technol.* **2021**, *36*, 115007. [CrossRef]
40. Yan, S.; Shi-Jin, D. Effects of Hydrogen Impurities on Performances and Electrical Reliabilities of Indium-Gallium-Zinc Oxide Thin Film Transistors. *Acta Phys. Sin.* **2017**, *66*, 218502.

41. Lin, D.; Zheng, X.; Yang, J.; Li, K.; Shao, J.; Zhang, Q. Annealing Effects on the Performances of Bismuth-Doped Indium Zinc Oxide Thin-Film Transistors. *J. Mater. Sci. Mater. Electron.* **2019**, *30*, 12929–12936. [CrossRef]
42. Sung, T.; Song, M.-K.; Jung, S.-Y.; Lee, S.; Song, Y.-W.; Park, S.; Kwon, J.-Y. Vacuum-Free Solution-Based Metallization (VSM) of a-IGZO Using Trimethylaluminium Solution. *RSC Adv.* **2022**, *12*, 3518–3523. [CrossRef] [PubMed]
43. Abliz, A.; Gao, Q.; Wan, D.; Liu, X.; Xu, L.; Liu, C.; Jiang, C.; Li, X.; Chen, H.; Guo, T. Effects of Nitrogen and Hydrogen Codoping on the Electrical Performance and Reliability of InGaZnO Thin-Film Transistors. *ACS Appl. Mater. Interfaces* **2017**, *9*, 10798–10804. [CrossRef] [PubMed]

Disclaimer/Publisher's Note: The statements, opinions and data contained in all publications are solely those of the individual author(s) and contributor(s) and not of MDPI and/or the editor(s). MDPI and/or the editor(s) disclaim responsibility for any injury to people or property resulting from any ideas, methods, instructions or products referred to in the content.

micromachines

MDPI

Communication

Optimization of Pulsed Laser Energy Density for the Preparation of MoS_2 Film and Its Device by Pulsed Laser Deposition

Wei Cai [1], Yuxiang Liu [2], Rihui Yao [2], Weijian Yuan [2], Honglong Ning [2,*], Yucheng Huang [2], Shaojie Jin [2], Xuecong Fang [2], Ruhai Guo [3,*] and Junbiao Peng [2]

1 Ji Hua Laboratory, Foshan 528000, China; caiwei@jihualab.ac.cn
2 Guangdong Basic Research Center of Excellence for Energy & Information Polymer Materials, State Key Laboratory of Luminescent Materials and Devices, School of Materials Sciences and Engineering, South China University of Technology, Guangzhou 510640, China; lyx19924688627@163.com (Y.L.); yaorihui@scut.edu.cn (R.Y.); wjyuan163@163.com (W.Y.); yucheng_h@163.com (Y.H.); 18770602788@163.com (S.J.); fxc20020122@163.com (X.F.); psjbpeng@scut.edu.cn (J.P.)
3 GDJH Advanced Display Equipment Ltd., Foshan 528000, China
* Correspondence: ninghl@scut.edu.cn (H.N.); hitgrh@163.com (R.G.)

Abstract: This article aims to explore the most optimal pulsed laser energy density when using the pulsed laser deposition (PLD) process to prepare the MoS_2 films. We gradually increased the pulsed laser energy density from 70 $mJ \cdot cm^{-2}$ to 110 $mJ \cdot cm^{-2}$ and finally determined that 100 $mJ \cdot cm^{-2}$ was the best-pulsed laser energy density for MoS_2 films by PLD. The surface morphology and crystallization of the MoS_2 films prepared under this condition are the best. The films consist of a high-crystallized 2H-MoS_2 phase with strong (002) preferential orientation, and their direct optical band gap (E_g) is 1.614 eV. At the same time, the Si/MoS_2 heterojunction prepared under the optimal pulsed laser energy density shows an opening voltage of 0.61 V and a rectification ratio of 457.0.

Keywords: MoS_2; pulsed laser deposition; pulsed laser energy density; heterojunction

check for updates

Citation: Cai, W.; Liu, Y.; Yao, R.; Yuan, W.; Ning, H.; Huang, Y.; Jin, S.; Fang, X.; Guo, R.; Peng, J. Optimization of Pulsed Laser Energy Density for the Preparation of MoS_2 Film and Its Device by Pulsed Laser Deposition. *Micromachines* **2024**, *15*, 945. https://doi.org/10.3390/mi15080945

Academic Editor: Chengyuan Dong

Received: 21 June 2024
Revised: 22 July 2024
Accepted: 23 July 2024
Published: 24 July 2024

Copyright: © 2024 by the authors. Licensee MDPI, Basel, Switzerland. This article is an open access article distributed under the terms and conditions of the Creative Commons Attribution (CC BY) license (https://creativecommons.org/licenses/by/4.0/).

1. Introduction

In recent years, two-dimensional materials such as graphene, transition metal sulfide (TMDCs), black phosphorus, and hexagonal boron nitride (h-BN) have been widely used in the field of optoelectronics [1–4]. The typical TMDCs material MoS_2 is one of the most commonly used and promising two-dimensional semiconductor materials. With its adjustable optical band gap and unique photoelectric characteristics, MoS_2 has shown promising application prospects in the fields of transistors, photoelectronic devices, and energy storage devices [5–7].

Given the excellent properties of MoS_2, researchers have developed several methods for preparing MoS_2 films of different thicknesses. The mechanical stripping method uses a special tape to overcome the van der Waals force between MoS_2 layers repeatedly peeling off the entire material, and finally obtaining a single layer to hundreds of layers of MoS_2 nanosheets [8]. The method is simple to operate and has a high stripping speed. However, the shape and thickness of the peeled nanosheets are uncontrollable, the efficiency of this preparation is low and the repeatability is poor. Therefore, this method is mainly used in the laboratory [9]. Chemical vapor deposition (CVD) is another method that is relatively easy to achieve large-scale production of MoS_2 films. This method uses transition metal (Mo) or transition metal oxide (MoO_3) to prepare MoS_2 by oxidation–reduction reaction with sulfur (S) [10–12]. However, CVD still has challenging problems in terms of thickness control, purity, and structure/morphology uniformity [13]. Therefore, it is currently necessary to find a method that can produce large-area MoS_2 films at relatively low temperatures [14–16].

Pulsed laser deposition (PLD) is a technology that uses pulsed laser to bombard the target and deposit the bombarded plasma on the substrate for film growth. PLD is known for its strong process latitude. By accurately and independently controlling the growth process parameters such as the frequency and energy of the pulsed laser and background pressure, it is possible to produce high-quality MoS_2 films from a few nanometers to a few microns in thickness. By accurately and independently controlling the growth process parameters such as the frequency and energy of the pulsed laser and background pressure, it is possible to produce high-quality MoS_2 films from a few nanometers to a few microns in thickness. Tumino [17] et al. studied the surface properties of single-layer MoS_2 deposited on Au (111) by PLD, starting from the growth of MoS_2 nanocrystals to the formation of a single-layer film uniformly covering the substrate surface on the centimeter scale. Siegel et al. [18] have demonstrated the growth of high-quality monolayer and few-layer MoS_2 films on transparent sapphire substrates using a PLD technique. The number of MoS_2 monolayers in the films was very precisely controlled by varying the number of pulsed laser pulses. Serna et al. [19] found a scalable and catalyst-free method to deposit MoS_2 films over large areas on a wide range of substrates without any additional surface preparation, including single-crystal (sapphire and quartz), polycrystalline (HfO_2), and amorphous (SiO_2) substrates. Tumino and Siegel successfully achieved the precise control of MoS_2 films' thickness on the gold substrate or transparent sapphire substrates by controlling the process parameters of PLD, but the studies about MoS_2 films deposited on the silicon substrates by PLD are few relatively.

Apart from the thickness of MoS_2 film, there is no clear demonstration of the relationship between the properties of the film and the process parameters, but this one is actually crucial. Increasing the energy density of the pulsed laser usually results in more target materials being vaporized and increasing the growth rate of the film. However, high energy density may lead to the spatter of large particles and damage the uniformity and smoothness of the films. The lower frequency of the pulsed laser slows down the growth rate of the films, which is favorable for improving the crystal quality, but the growth time will be prolonged. Higher substrate temperature helps to increase the surface mobility of the deposited atoms and promotes crystal growth, which improves the crystallinity and orientation of the film.

There are many controllable process parameters for films' quality, and the strategy of preparing MoS_2 films by PLD still needs further research. Thus, a systematic analysis of the process parameters for PLD depositing MoS_2 films is required. The effect of varying a process parameter during the deposition process on the properties of films and devices is determined by the control variable method. Since the laser energy density mainly determines the generation of the plume and the deposition process, this work chooses to find the relationship between laser energy density and the properties of films and devices. In this paper, we have been able to identify an optimal pulsed laser energy density (W_{PL}) for preparing highly crystallized MoS_2 films exhibiting a direct bandgap of 1.614 eV and a very high photoelectric performance.

2. Results and Discussion

We used PLD to prepare different MoS_2 films on the silicon substrate by setting W_{PL} = 70 mJ·cm^{-2}, 80 mJ·cm^{-2}, 90 mJ·cm^{-2}, 100 mJ·cm^{-2} and 110 mJ·cm^{-2}, respectively. All other process parameters were consistent (the distance between the target and the substrate was 7.5 cm, the deposition pressure was 5×10^{-4} Pa, the pulsed laser frequency was 5 Hz, the number of laser pulses was 3000 and the substrate temperature was 400 °C).

Figure 1 shows the surface morphology of MoS_2 films under a scanning electron microscope (SEM). With W_{PL} increasing, the uniformity of films gradually improved and the irregular bulk particles on the surface gradually disappeared. The thickness and density of the films were measured using X-ray reflectivity (XRR) and the roughness of the film's surface was measured using an atomic force microscope (AFM), the data are shown in Table 1. The thickness of films increased with laser energy density increasing from

70 mJ·cm^{-2} to 100 mJ·cm^{-2}. However, the films' thickness decreased to 9.938 nm sharply and was smaller than the films with W_{PL} = 70 mJ·cm^{-2} when the W_{PL} was 110 mJ·cm^{-2}. The higher W_{PL} on the MoS_2 target leads to a stronger bombardment effect, the escaped atoms from the target's surface have higher energy, and the collision effect between these atoms is more intense when atoms reach the substrate. When the energy is too high (110 mJ·cm^{-2}), the escaping atoms have so high kinetic energy that knocks away the films already deposited on the substrate, so the thickness of the films decreases dramatically, and the roughness becomes larger.

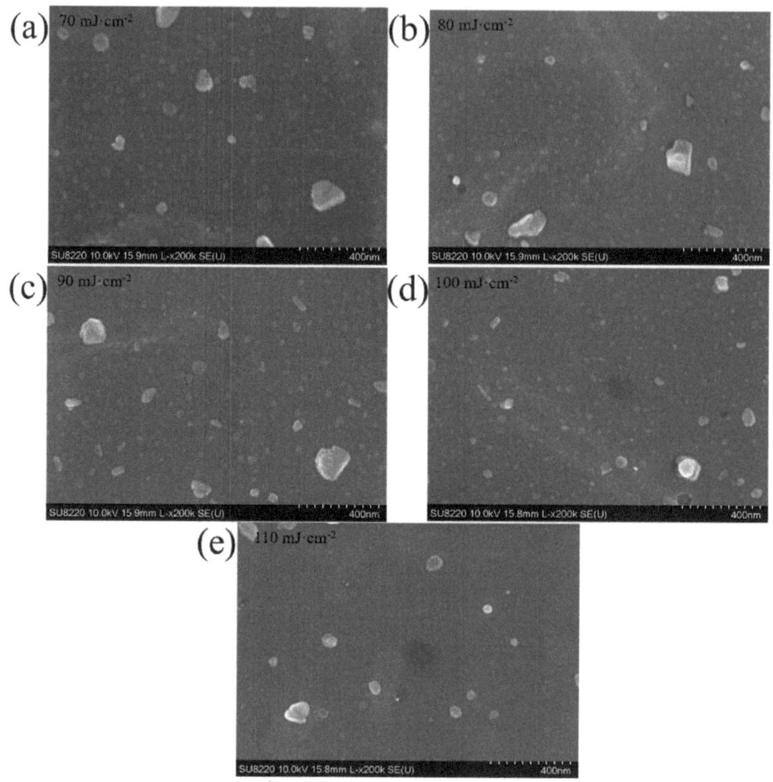

Figure 1. SEM images of MoS_2 films deposited with different W_{PL} ((**a**) 70 mJ·cm^{-2}, (**b**) 80 mJ·cm^{-2}, (**c**) 90 mJ·cm^{-2}, (**d**) 100 mJ·cm^{-2} and (**e**) 110 mJ·cm^{-2}).

Table 1. Thickness, density, and roughness of MoS_2 films deposited with different laser energy densities.

Energy (mJ·cm^{-2})	Thickness (nm)	Density (g·cm^{-3})	Roughness (nm)
70	12.984	4.971	3.74
80	12.662	5.173	1.8
90	13.821	5.126	2.9
100	14.026	5.143	1.44
110	9.938	5.210	1.89

Thus, the W_{PL} should be chosen under 110 mJ·cm^{-2}. In addition, it can be seen from Table 1 that the films prepared at 100 mJ·cm^{-2} have the smallest roughness, indicating that the film's surface morphology is the best, and thus 100 mJ·cm^{-2} is a good choice.

X-ray diffraction (XRD) test was performed on MoS_2 films deposited with different W_{PL} to characterize their crystallization and the results are shown in Figure 2a. All kinds of MoS_2 films have sharp diffraction peaks at $2\theta = 13.66°$, corresponding to the (002) crystal face of the hexagonal (2H) phase MoS_2. When W_{PL} increased from 70 mJ·cm^{-2} to 100 mJ·cm^{-2}, the intensity of the diffraction peak increased gradually. However, when W_{PL} increased to 110 mJ·cm^{-2}, the intensity of the diffraction peak decreased. This shows that when W_{PL} is 100 mJ·cm^{-2}, the crystallization of MoS_2 films is best.

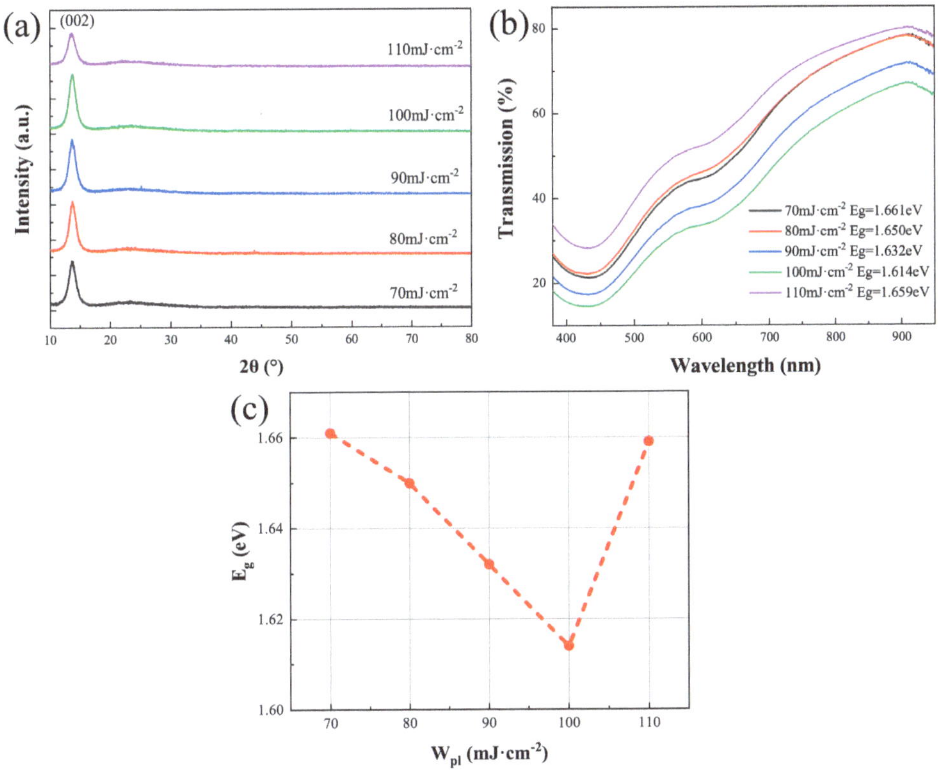

Figure 2. (**a**) XRD patterns of MoS_2 films deposited with different W_{PL}. (**b**) Optical transmission spectrum of MoS_2 films deposited with different W_{PL} in visible region. (**c**) Calculated E_g of MoS_2 films deposited with different W_{PL}.

Figure 2b shows the optical transmission spectrum of MoS_2 films deposited on the quartz substrates with different W_{PL} in the visible region. The spectrum clearly shows that MoS_2 films absorb more light at shorter wavelengths and have the strongest absorption waves at 430 nm. Combining the transmission spectrum and Tauc equation ($(\alpha h v)^{\frac{1}{m}} = B(hv - E_g)$, where α is the absorption coefficient, m is directly related to the semiconductor type, generally using m = 1/2 in calculating direct band gap), we calculated the direct optical band gap (E_g) of MoS_2 films as shown in Figure 2c. As W_{PL} increased from 70 mJ·cm^{-2} to 100 mJ·cm^{-2}, E_g decreased significantly from 1.661 eV to 1.614 eV. When W_{PL} increased to 110 mJ·cm^{-2} E_g again increased to 1.659 eV. This E_g variation is definitely dictated by the crystallinity and most importantly the nanostructural arrangement of the MoS_2 films, which is consistent with the above XRD patterns. The E_g has a great influence on the optical properties of materials. In general, materials with smaller E_g have a wider absorption range for light waves. Therefore, the MoS_2 films prepared by W_{PL} = 100 mJ·cm^{-2} are more advantageous in the application of optoelectronic devices.

The above test results show that W_{PL} = 100 mJ·cm^{-2} is the most suitable setting for preparing MoS_2 films by PLD. Therefore, we used PLD to deposit MoS_2 film and a gold electrode on Si substrates to obtain Si/MoS_2 heterojunctions. The structure of the heterojunctions is shown in Figure 3a. By setting W_{PL} = 100 mJ·cm^{-2} and changing the number of laser pulses (1000, 3000, 5000), we could obtain MoS_2 films with different thicknesses. According to the different thicknesses of deposited MoS_2 films (2.0 nm, 4.1 nm and 4.7 nm), the heterojunctions were named Si/MoS_2-1, Si/MoS-2, and Si/MoS_2-3, respectively.

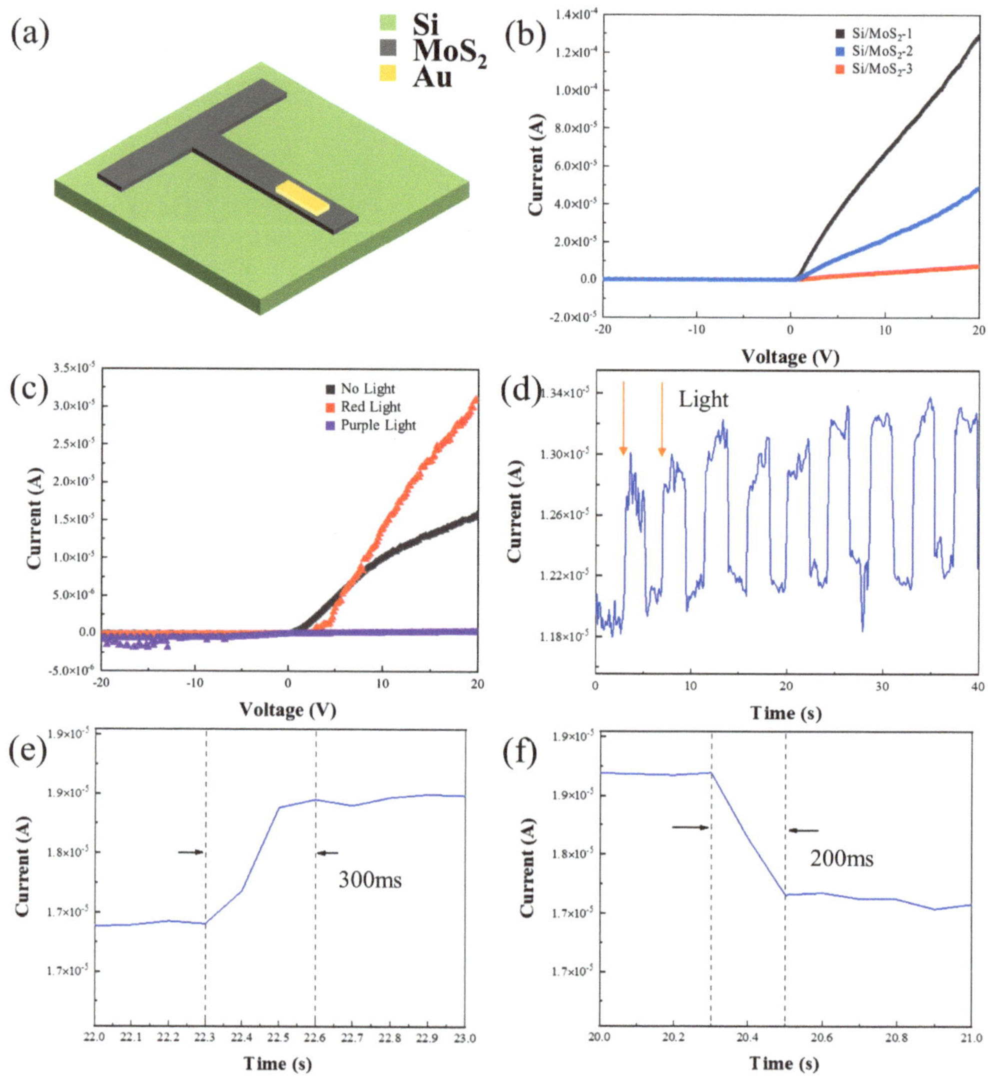

Figure 3. (**a**) The structure of Si/MoS_2 heterojunctions. (**b**) The I-V characteristic of Si/MoS_2 heterojunctions with different thicknesses tested in dark conditions. (**c**) The I-V characteristic of Si/MoS_2 heterojunctions under different lights. (**d**) The cyclic optical response curve of Si/MoS_2 heterojunctions (**e**) The optical response time test curves of Si/MoS_2 heterojunctions. (**f**) The optical recovery time test curves of Si/MoS_2 heterojunctions.

The electrical properties of Si/MoS_2 heterojunctions were studied under dark conditions and the test results are shown in Figure 3b. It can be observed that the Si/MoS_2 heterojunctions possess a characteristic of unilateral conductivity. The opening voltage (V_{on}), forward current (I_d^+), reverse current (I_d^-), and rectification ratio (A) data of three kinds of samples are listed in Table 2. The A of Si/MoS_2-3 is only 67.7 and its electrical performance is poor. The A of Si/MoS_2-1 and Si/MoS_2-2 is 457.0 and 467.6, respectively. The V_{on} of Si/MoS_2-1 is 0.61 V, which is smaller than that of Si/MoS_2-2, so the electrical performance of Si/MoS_2-1 is relatively better. The reason why MoS_2 films show P-type semiconductor properties is the generation of S-vacancy. Therefore, under certain conditions, the more carriers the films can provide with more defects, the better the conductivity of the device. The films of Si/MoS_2-1 are the thinnest and have more defects, which shows the best electrical properties.

Table 2. The opening voltage (V_{on}), forward current (I_d^+), reverse current (I_d^-), and rectification ratio (A) of Si/MoS_2 heterojunctions with different thicknesses tested in dark conditions.

Sample	V_{on} (V)	I_d^+ (A)	I_d^- (A)	A
Si/MoS_2-1	0.61	4.89×10^{-5}	-1.07×10^{-7}	457.0
Si/MoS_2-2	0.66	1.30×10^{-4}	-2.78×10^{-8}	467.6
Si/MoS_2-3	0.54	7.38×10^{-6}	-1.09×10^{-7}	67.7

Subsequently, we tested the photoelectric performance of the Si/MoS_2 heterojunctions, and the results are shown in Figure 3c. It can be clearly observed that the response of Si/MoS_2 heterojunctions under red light and violet light was different from that under dark conditions. The reason is that the charge separation phenomenon occurred in Si/MoS_2 heterojunctions and resulted in photogenerated charge carriers under light. Compared with dark conditions, I_d^+, I_d^-, and V_{on} of heterojunctions increased significantly under red light conditions. This is because the photogenerated carriers produced after illumination increased the diffusion current concentration, breaking the balance between multiphoton diffusion and minority drift, and the drift current concentration began to increase, and the space charge region became wider. At this point, higher voltage was required to fill the increased portion of the space charge region, so V_{on} increased. When V = −20 V, I_d^- = -1.55×10^{-7} A. When V = 20 V, I_d^+ = 3.09×10^{-5} A. The V_{on} of Si/MoS_2 heterojunctions increased to 3.49 V, and the A reduced to 199.35. When exposed to violet light, the Si/MoS_2 heterojunctions lost their characteristic electrical properties. The reason may be that the energy of violet light is so high that the deep-level impurities became charged centers after ionization, scattering the carriers and reducing the carrier mobility and conductive performance.

Figure 3d shows the cyclic optical response curve of Si/MoS_2 heterojunctions. When red light was used, the Si/MoS_2 heterojunctions current increased, and the circulation still remained in the relative range after being repeated several times, which indicated the Si/MoS_2 heterojunctions possessed good repeatability. Figure 3e,f are the optical response time and recover time test curves of n-Si/MoS_2 heterojunctions. It can be seen that the response time was 300 ms and the recovery time was 200 ms, so the response speed is relatively rapid.

3. Conclusions

In conclusion, we have found the relationship between laser energy density and film properties when preparing MoS_2 films using PLD technology. When the laser energy density increased within a certain range, the thickness of the MoS_2 films increased and the optical band gap decreased gradually. When the laser energy density continued to increase until exceeding the optimal laser energy density, the thickness of the MoS_2 films decreased dramatically and the properties deteriorated severely. Therefore, we confirmed that the optimal laser energy density for the preparation of MoS_2 films using PLD technology is

100.0 mJ·cm^{-2}. The surface shape and crystallization of the MoS_2 films prepared under this condition are the best. Because the films consist of a high-crystallized 2H-MoS_2 phase with strong (002) preferential orientation, their direct optical band gap (E_g) is 1.614 eV and demonstrates excellent photoelectrical properties. At the same time, the Si/MoS_2 heterojunction prepared under the optimal pulsed laser energy density shows an opening voltage of 0.61 V and a rectification ratio of 457.0.

4. Experiment Section

(1) Target and substrates: The MoS_2 target used in this experiment was purchased from ZNXC Technology Co., Ltd. (Beijing, China). with a target size of 25.4 mm in diameter and 3 mm in thickness, and a purity of 99.99%. The substrates used in this experiment were silicon and quartz substrates, which were purchased from Guangzhou New Vision Optoelectronics Technology Co., Ltd. (Guangzhou, China).

(2) Substrates' cleaning: Firstly, the silicon and quartz substrates were discharged on a washing rack, and then they were ultrasonicated with isopropyl alcohol for 10 min, deionized water for 10 min each twice, and isopropyl alcohol for 10 min, and finally the cleaned substrates were placed in an oven at 80 °C for drying, and then taken out for use after 1 h.

(3) MoS_2 film deposition: The laser used in this study was a COMPex201 excimer laser manufactured by Coherent (Saxonburg, PA, USA), with KrF gas as the working gas and an intracavity working air pressure of 3400 mbar. The main wavelength of the output laser was 248 nm, the adjustable range of the pulsed laser frequency was from 1 to 10 Hz and the adjustable range of the laser pulse energy was from 180 to 750 mJ. MoS_2 target and substrates were placed in the PLD cavity of the No. 1 target position and the sample stage, respectively, the distance between the target and the substrate was 7.5 cm, the deposition pressure was 5×10^{-4} Pa, the laser frequency was 5 Hz, the number of laser pulses was 3000 times, the substrates' temperature was 400 °C. During the deposition process, the MoS_2 target was rotated while the laser beam was laterally swept across its entire surface to ensure a uniform erosion pattern of the target. This enables a more spatially extended source of ablated species covering more uniformly the substrate holder. Moreover, the substrate holder itself was concomitantly rotated to improve further the thickness uniformity of the films. Prior to each deposition, the MoS_2 target surface was in situ cleaned by ablating its surface for 5 min. The laser energy density was variable and was set to 70 mJ·cm^{-2}, 80 mJ·cm^{-2}, 90 mJ·cm^{-2}, 100 mJ·cm^{-2} and 110 mJ·cm^{-2}, respectively.

(4) Test and characterization: In this study, an X-ray diffractometer is mainly used to obtain the films' information such as thickness and crystallization. The X-ray diffractometer is Empyrean DY1577 (XRD: Malvern Panalytical Company, Almelo, The Netherlands), with a test voltage of 40 kV, a test current of 40 mA, a test angle of 2θ of 10~80°, a scanning step size of 0.02°, and a dwell time of 1.5 s per step. The transmission spectrum of the films was tested by UV–Vis spectrophotometer, and then the optical bandgap of the films was calculated by extrapolation. The equipment used in this study was a model UV-2600 spectrophotometer from Shimadzu Instruments Ltd. (Columbia, MD, USA). A semiconductor analyzer is used to test and characterize the electrical properties of the heterojunction. The equipment is Agilent 4155C (Agilent Company, Santa Clara, CA, USA), and by measuring the I-V curve of the heterojunction, turn-on voltage, rectification ratio, and reverse current can be obtained to measure the performance of the device.

Author Contributions: Conceptualization, W.C. and W.Y.; methodology, Y.L.; validation, S.J. and X.F.; investigation, Y.H.; resources, R.Y.; data curation, W.C.; writing—original draft preparation, Y.L.; writing—review and editing, W.C. and H.N.; supervision, H.N.; project administration, R.G.; funding acquisition, H.N. and J.P. All authors have read and agreed to the published version of the manuscript.

Funding: This research was funded by National Key R&D Program of China: No. 2022YFB3603600; National Natural Science Foundation of China: No. 62174057 and 22090024; Guangdong Natural Science Foundation: No. 2024A1515012216 and 2023A1515011026; Educational Commission of Guangdong Province: No. 2022ZDZX1002; Key R&D Plan of Guangdong Province: 2022B0303010001; State Key Lab of Luminescent Materials and Devices (Skllmd-2024-05), the open research fund od Songshan Lake Materials Laboratory (2022SLABFN05).

Data Availability Statement: All the relevant data are included in this published article.

Conflicts of Interest: The authors declare no conflicts of interest. Ruhai Guo is an employee of GDJH Advanced Display Equipment Ltd. The paper reflects the views of the scientists and not the company.

References

1. Li, S.; Huang, G.; Jia, Y.; Wang, B.; Wang, H.; Zhang, H. Photoelectronic Properties and Devices of 2D Xenes. *J. Mater. Sci. Technol.* **2022**, *126*, 44–59. [CrossRef]
2. Guo, L.; Wu, M.; Cao, T.; Monahan, D.M.; Lee, Y.-H.; Louie, S.G.; Fleming, G.R. Exchange-Driven Intravalley Mixing of Excitons in Monolayer Transition Metal Dichalcogenides. *Nat. Phys.* **2019**, *15*, 228–232. [CrossRef]
3. Li, L.; Yu, Y.; Ye, G.J.; Ge, Q.; Ou, X.; Wu, H.; Feng, D.; Chen, X.H.; Zhang, Y. Black Phosphorus Field-Effect Transistors. *Nat. Nanotechnol.* **2014**, *9*, 372–377. [CrossRef] [PubMed]
4. Liu, Y.; Ong, Z.-Y.; Wu, J.; Zhao, Y.; Watanabe, K.; Taniguchi, T.; Chi, D.; Zhang, G.; Thong, J.T.L.; Qiu, C.-W.; et al. Thermal Conductance of the 2D MoS_2/h-BN and Graphene/h-BN Interfaces. *Sci. Rep.* **2017**, *7*, 43886. [CrossRef] [PubMed]
5. Chen, Y.; Sun, M. Two-Dimensional WS_2/MoS_2 Heterostructures: Properties and Applications. *Nanoscale* **2021**, *13*, 5594–5619. [CrossRef] [PubMed]
6. Lu, Y.; Yang, G.; Wang, F.; Lu, N. Enhanced Photoresponse of Monolayer Molybdenum Disulfide (MoS_2) Based on Microcavity Structure. *Superlattices Microstruct.* **2018**, *117*, 163–168. [CrossRef]
7. Ma, Y.; Liu, J.; Lin, Y.; Jia, Y. Recent Advances in Hierarchical MoS_2/Graphene-Based Materials for Supercapacitor Applications. *Phys. Chem. Chem. Phys.* **2023**, *25*, 8263–8280. [CrossRef] [PubMed]
8. Lopez-Sanchez, O.; Lembke, D.; Kayci, M.; Radenovic, A.; Kis, A. Ultrasensitive Photodetectors Based on Monolayer MoS_2. *Nat. Nanotechnol.* **2013**, *8*, 497–501. [CrossRef]
9. Kim, S.; Park, W.; Kim, D.; Kang, J.; Lee, J.; Jang, H.Y.; Song, S.H.; Cho, B.; Lee, D. Novel Exfoliation of High-Quality 2H-MoS_2 Nanoflakes for Solution-Processed Photodetector. *Nanomaterials* **2020**, *10*, 1045. [CrossRef] [PubMed]
10. Jian, J.; Chang, H.; Dong, P.; Bai, Z.; Zuo, K. A Mechanism for the Variation in the Photoelectric Performance of a Photodetector Based on CVD-Grown 2D MoS_2. *RSC Adv.* **2021**, *11*, 5204–5217. [PubMed]
11. Zang, L.; Chen, L.; Tan, D.; Cao, X.; Sun, N.; Jiang, C. Research on Multi-Morphology Evolution of MoS_2 in Chemical Vapor Deposition. *ChemistrySelect* **2021**, *6*, 8107–8113. [CrossRef]
12. Gao, Q.; Chen, L.; Chen, S.; Zhang, Z.; Yang, J.; Pan, X.; Yi, Z.; Liu, L.; Chi, F.; Liu, P.; et al. NaCl-Assisted Chemical Vapor Deposition of Large-Domain Bilayer MoS_2 on Soda-Lime Glass. *Nanomaterials* **2022**, *12*, 2913. [CrossRef] [PubMed]
13. Mouloua, D.; LeBlanc-Lavoie, J.; Pichon, L.; Rajput, N.S.; El Marssi, M.; Jouiad, M.; El Khakani, M.A. Tuning the Optoelectronic Properties of Pulsed Laser Deposited "3D"-MoS_2 Films via the Degree of Vertical Alignment of Their Constituting Layers. *Adv. Opt. Mater.* **2024**, *12*, 2302966. [CrossRef]
14. Chen, J.; Yang, C.; Xu, J.; Gao, L.; Guo, T.; Jia, S.; Zhang, P.; Xiao, Y.; Chen, J.; Zhao, Y.; et al. The Plasticity of Synaptic Memristor Based on 2D-MoS_2 Thin Film Prepared in Large-Scale by a PLD-Assisted CVD Method. *Mater. Today Commun.* **2023**, *35*, 105511. [CrossRef]
15. Loh, T.A.J.; Chua, D.H.C. Growth Mechanism of Pulsed Laser Fabricated Few-Layer MoS_2 on Metal Substrates. *ACS Appl. Mater. Interfaces* **2014**, *6*, 15966–15971. [CrossRef]
16. Yu, J.; Han, W.; Suleiman, A.A.; Han, S.; Miao, N.; Ling, F.C.-C. Recent Advances on Pulsed Laser Deposition of Large-Scale Thin Films. *Small Methods* **2023**, *8*, 2301282. [CrossRef]
17. Tumino, F.; Casari, C.S.; Passoni, M.; Russo, V.; Bassi, A.L. Pulsed Laser Deposition of Single-Layer MoS_2 on Au (111): From Nanosized Crystals to Large-Area Films. *Nanoscale Adv.* **2019**, *1*, 643–655. [CrossRef] [PubMed]
18. Siegel, G.; Venkata Subbaiah, Y.P.; Prestgard, M.C.; Tiwari, A. Growth of Centimeter-Scale Atomically Thin MoS_2 Films by Pulsed Laser Deposition. *APL Mater.* **2015**, *3*, 056103. [CrossRef]
19. Serna, M.I.; Yoo, S.H.; Moreno, S.; Xi, Y.; Oviedo, J.P.; Choi, H.; Alshareef, H.N.; Kim, M.J.; Minary-Jolandan, M.; Quevedo-Lopez, M.A. Large-Area Deposition of MoS_2 by Pulsed Laser Deposition with In Situ Thickness Control. *ACS Nano* **2016**, *10*, 6054–6061. [CrossRef] [PubMed]

Disclaimer/Publisher's Note: The statements, opinions and data contained in all publications are solely those of the individual author(s) and contributor(s) and not of MDPI and/or the editor(s). MDPI and/or the editor(s) disclaim responsibility for any injury to people or property resulting from any ideas, methods, instructions or products referred to in the content.

micromachines

Article

Polarization-Insensitive Lithium Niobate-on-Insulator Interferometer

Jiali Liao [1,2,*], Linke Liu [1], Yanling Sun [1,*], Zihao Wang [1], Wei Li [1], Jinrong Lan [1], Lin Ma [1] and Zhenzhong Lu [1]

[1] School of Optoelectronic Engineering, Xidian University, Xi'an 710071, China; 22191215011@stu.xidian.edu.cn (L.L.); wangzihao@xidian.edu.cn (Z.W.); wli0805@stu.xidian.edu.cn (W.L.); 21051212315@stu.xidian.edu.cn (J.L.); lma@mail.xidian.edu.cn (L.M.); zzluxidian@126.com (Z.L.)
[2] Science and Technology on Electromechanical Dynamic Control Laboratory, Xi'an 710065, China
* Correspondence: liaojiali@xidian.edu.cn (J.L.); ylsun@mail.xidian.edu.cn (Y.S.)

Abstract: The key components of a polarization-independent electro-optic (EO) interferometer, including the beam splitter, mode converter, and directional coupler, are designed based on a lithium niobate (LN) platform on an integrated insulator to ensure high extinction ratios. By elaborately designing the geometric structure of the multimode interference (MMI) coupler, beam splitting of equal proportions and directional coupling of higher-order modes are realized. The most prominent characteristic of the proposed interferometer is polarization insensitivity, which is realized by converting TM polarization into TE polarization using a mode converter, providing conditions for the study of light with different polarizations. At 1550 nm, the visibility of the interferometer is 97.59% and 98.16% for TE and TM, respectively, demonstrating the high performance of the proposed LN polarization-independent interferometer.

Keywords: electro-optic modulation; polarization independent; lithium niobate; mode converter

check for updates

Citation: Liao, J.; Liu, L.; Sun, Y.; Wang, Z.; Li, W.; Lan, J.; Ma, L.; Lu, Z. Polarization-Insensitive Lithium Niobate-on-Insulator Interferometer. *Micromachines* **2024**, *15*, 983. https://doi.org/10.3390/mi15080983

Academic Editor: Chengyuan Dong

Received: 22 June 2024
Revised: 20 July 2024
Accepted: 28 July 2024
Published: 30 July 2024

Copyright: © 2024 by the authors. Licensee MDPI, Basel, Switzerland. This article is an open access article distributed under the terms and conditions of the Creative Commons Attribution (CC BY) license (https://creativecommons.org/licenses/by/4.0/).

1. Introduction

Over the past decade, there has been a series of significant research progresses in optical interconnections, which have gradually replaced copper interconnections due to their superior bandwidth [1–5]. One crucial component in this transmission is the EO modulator, responsible for converting electrical signals into optical signals [6–8]. Although many silicon EO modulators have been reported, it is believed that their current capabilities are approaching their physical limits, such as the modulation speed and optical loss, while future applications will require much higher performance for modulators.

Despite advancements in various materials [9–13], LN remains the preferred choice for EO modulators due to its remarkable linear EO effect, substantial EO coefficient, low optical absorption loss, and proven reliability [14,15]. Recently, the development of lithium niobate thin film on insulators has further improved LN modulators by offering a higher refractive-index contrast compared to traditional bulky LN modulators. One limitation of LN modulators is their polarization dependence, which results from the anisotropy of LN crystals. However, polarization-independent modulators are highly desired by many application fields, such as optical communication [4,16–19], optical signal processing, and microwave photonics [20,21]. A polarization-independent EO modulator with a loss of 0.04 dB was reported, but this modulator used Z-CUT LN, which could not use the high EO coefficient of X-CUT LN; therefore, the modulator needed high voltage [22]. An LN interferometer, which realized a polarization beam splitter and a polarization-insensitive switch by LN thermo-optic MZI, was reported in [23].

In this paper, we proposed a novel polarization-independent LN interferometer by designing a mode converter to convert TM to TE polarization before the EO modulation. The interferometer was designed based on an X-CUT LN crystal, which enabled effective modulation along the x-axis, leveraging the substantial EO coefficient of LN in this direction and aligning the crystal polarization with the light field in the waveguide. Apart from the mode converter, the functional components of the interferometer were well designed, including the LN waveguides and the 1 × 2 and 2 × 2 MMI. The performance of the designed polarization-independent interferometer was investigated by simulation. Despite the polarization, the input light was modulated with high extinction ratios, demonstrating the superior polarization independence of the proposed LN interferometer.

2. Design of the Polarization-Independent LN EO Interferometer

The schematic diagram of the proposed polarization-insensitive LN interferometer is shown in Figure 1. The light is injected from the input waveguide and divided into two modulation waveguides after passing through the optical beam splitter. The two mode converters are integrated at the front and the back of the modulation waveguides, respectively. After passing through the front mode converter, the TE_0 mode maintains its polarization without mode conversion and is modulated directly, while the TM_0 mode is converted to the TE_1 mode, guaranteeing the subsequent modulation. The TE_1 (converted from the TM_0 mode) and initial TE_0 mode are modulated by the electrical signal in the modulation regions, then the lights are converted to the same mode as the input lights, and finally the lights are combined in the 2 × 2 MMI. Meanwhile, the modulation regions use the reported structure with the voltage–length product of 2.2 Vcm and an insertion loss of 2.5 dB [24]. Due to limited equipment and computing power, this study only simulated each device separately and then evaluated the overall device after obtaining the results.

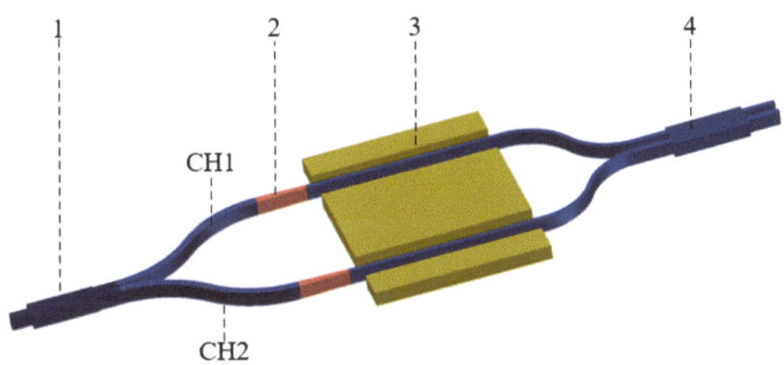

Figure 1. Schematic of the designed interferometer: 1: beam splitter; 2: mode converter; 3: electrode; 4: directional coupler.

2.1. Mode Converter

The mode converter is the key component of a polarization-independent interferometer. It was designed based on mode hybridization in the waveguide taper, with a variable waveguide width. The employed taper structure is shown in Figure 2a. This structure is a conventional lateral taper in which the waveguide width varies while maintaining a constant etch depth. Due to its simplicity in design and fabrication, regular lateral tapers are commonly used to adjust the lateral dimensions of waveguide modes [25]. In an LN waveguide, mode conversion between TM_0 and TE_1 modes, which is judged based on intensity distribution [26,27], is expected to occur as light propagates through the taper structure.

We analyzed the modes using the finite-difference eigenmode (FDE) method in the waveguide, with different widths between 0.5 μm and 3 μm used to characterize mode transmissions in tapered waveguides, and the results are shown in Figure 3.

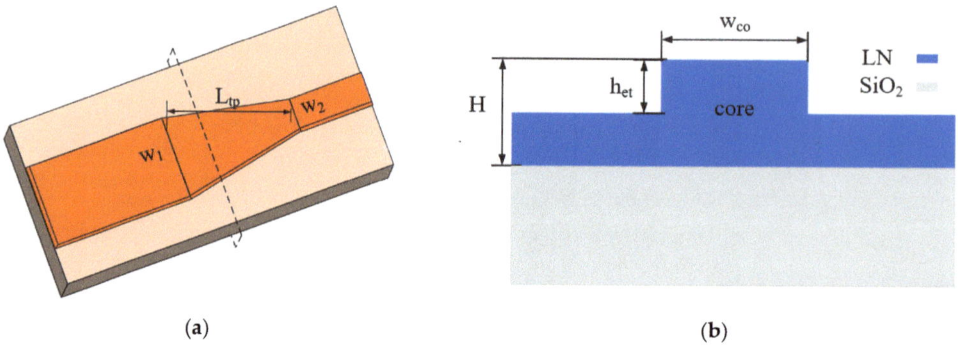

Figure 2. (**a**) Schematic diagram of mode converter. (**b**) Cross-section of the waveguide.

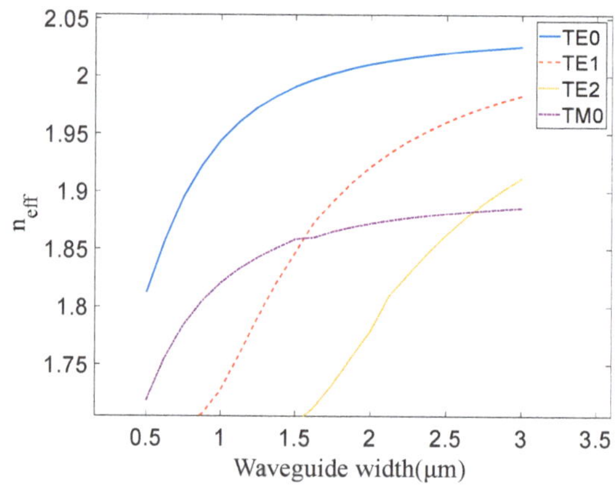

Figure 3. The calculated effective refractive index for the eigenmodes with different widths.

In Figure 3, the effective refractive index of the LN waveguide is shown for an etch depth (h_{et}) of 0.4 μm and a total height (H) of 0.6 μm while the core width (w_{co}) increases from 0.5 μm to 3 μm. Due to the vertical asymmetry of the waveguide, mode conversion phenomena were predicted to occur in certain width ranges. As shown in Figure 3, the effective refractive index curves of TM_0 and TE_1 intersect at approximately w_{co} = 1.6 μm, where mode conversion occurs. This conversion is achieved when the waveguide width (w_1, w_2) satisfies the condition $w_1 < w_{co} < w_2$, which is used to perform the polarization rotation. Therefore, the waveguide width of the taper region is chosen to decrease from 2 μm to 1.4 μm, and the corresponding mode conversion efficiency is estimated. As shown in Figure 4, using the eigenmode expansion (EME) method, high mode conversion efficiency (>90%) can be achieved when the length of the conversion region is longer than 125 μm. Mode conversion efficiency measures the ratio of the energy of the goal mode to the input mode. Since the mode converter is reversible, mode conversion efficiency refers to the efficiency of the conversion between TE_1 and TM_0 modes; meanwhile, the light mode can be judged by the intensity distribution.

Utilizing the 3D finite-difference time-domain (FDTD) method, the results of light mode conversion with TE_0, TM_0, and TE_1 input modes are shown in Figure 5, respectively. It shows that the input TE_0 mode does not undergo mode conversion, while the other two input modes experience a remarkable mode conversion.

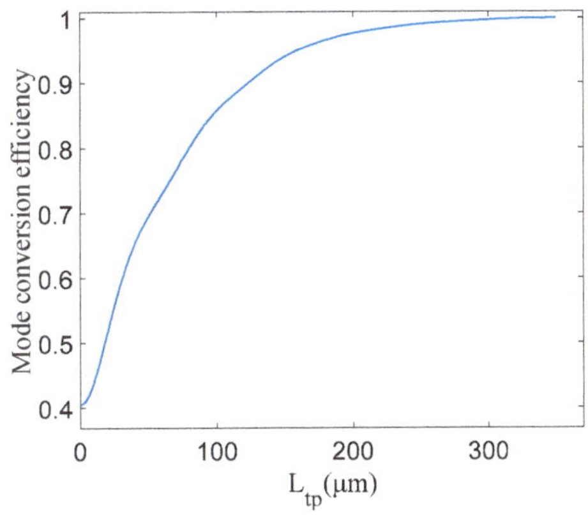

Figure 4. Relation between taper structure length and conversion efficiency.

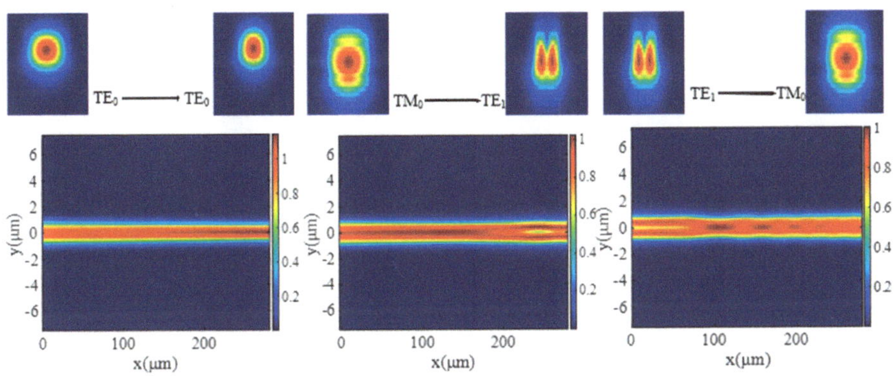

Figure 5. Schematics of TE_0-TE_0, TM_0-TE_1, and TE_1-TM_0 mode converters with corresponding mode profiles.

2.2. Multimode Interference Coupler

The operational principle of an MMI coupler is based on the self-imaging phenomenon in multimode waveguides. This phenomenon results in the periodic reproduction of the input field distribution as one or more images. Therefore, a fundamental prerequisite for achieving the MMI effects is a waveguide capable of supporting multiple modes. In the case of ridge waveguides, sufficiently wide waveguides can support a variety of modes. The wider the waveguide is, the larger the number of modes it can support, leading to the enhancement of image quality. However, it is worth noting that a wider waveguide requires a longer MMI region.

In the proposed interferometer, two kinds of MMIs are designed, i.e., a 1 × 2 MMI and a 2 × 2 MMI are designed at the input and output ends, respectively. The 1 × 2 MMI is used as a beam splitter device, while the 2 × 2 MMI is used as a directional coupler. The 1 × 2 MMI distributes the input light energy into both arms of the splitter equally, while the 2 × 2 MMI ensures uniform distribution of light input from one arm to both output arms in the directional coupler. By designing the width of the 2 × 2 MMI, MMIs can satisfy various requirements, for example, near-zero anomalous group-velocity dispersion for photon-pair generation, optical switch [1,10,28], and 3 dB beam splitter [29]. However, this study concerns the transmission of two polarizations.

Simulations of the transmission of TE_0 and TM_0 modes were conducted using the 3D FDTD method. Figure 6 shows the propagation of TE_0 and TM_0 modes in the 1 × 2 MMI and 2 × 2 MMI, respectively. It shows that the 1 × 2 MMI can evenly split the input light energy and transfer it to two output ports, whether it is the TE_0 or TM_0 mode. Furthermore, the input light from any input port of the 2 × 2 MMI can be evenly split into two output ports on the condition of TE_0 and TM_0 modes indicating that the 2 × 2 MMI can realize beam splitting for TE_0 and TM_0.

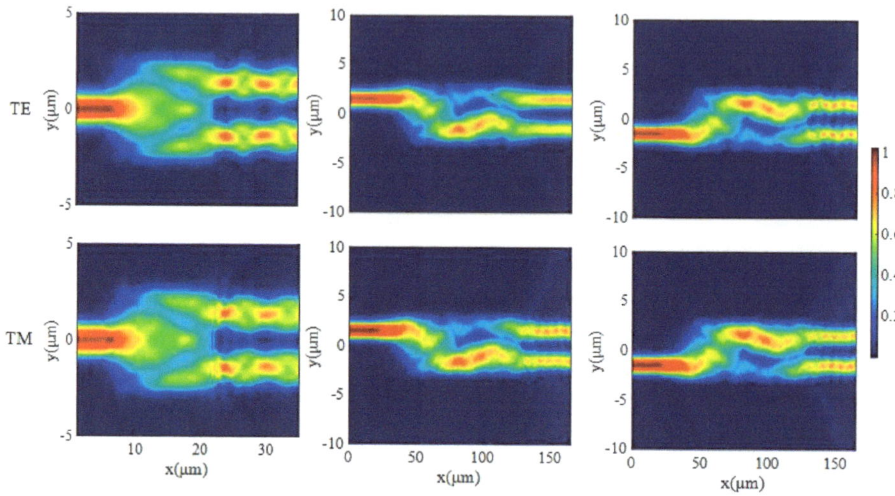

Figure 6. The 1 × 2 MMI and 2 × 2 MMI electric field evolution diagram.

Defining L_π as the beat length of the two lowest-order modes, it can be described by

$$L_\pi = \frac{\pi}{\beta_0 - \beta_1} = \frac{4 n_r {W_e}^2}{3\lambda_0} \tag{1}$$

where W_e is the effective waveguide width in the multimode interference region, n_r is the core refractive index, the incident wavelength in free space is λ_0, and the propagation constant of order v is β_v and can be described as

$$\beta_v = \frac{2\pi n_v}{\lambda} \tag{2}$$

where n_v is the effective refractive index of the v mode.

According to different incident conditions, an MMI can be divided into three types: ordinary interference, symmetrical interference, and paired interference. The three types have different positions for the first N-fold, which are $3L_\pi/N$, $3L_\pi/4N$, and L_π/N, respectively.

In order to reduce loss, a conical waveguide is used at the connection of the multimode waveguide region. Additionally, the 1 × 2 MMI is selected to be the symmetric interference type while the 2 × 2 MMI is selected to be the ordinary interference type to reduce the footprint of the interferometers.

The 1 × 2 MMI multimode waveguide width is preliminarily set to be 5.2 μm. The shortest length of the 1 × 2 MMI coupler is $L_{MMI} = 3L_\pi/8$. When $W_{MMI} = 5.2$ μm, n_0 and n_1 are found to be 2.035 and 2.021 by calculating that β_0 and β_1 are 8.251 and 8.194, respectively. The length of the L_π is 55.11 μm and the length of the MMI is calculated as $L_{MMI} = 20.67$ μm. It is a rough initial value, and it is necessary to scan around the value to determine the exact value.

We utilize the 3D FDTD method to simulate light evolution in the MMI with different widths and lengths. As shown in Figure 7a, when the length of the MMI is 17 μm, the width of the MMI is 5.2 μm, and the gap width is 0.5 μm, the transmissions of both modes reach a maximum, which is more than 94.45%. When the length is between 14 and 22 μm, the MMI has a good transmission result. As shown in Figure 7b, the transmission of the MMI was calculated with the width varying from 4.5 μm to 6 μm, with the length of MMI as 17 μm, and the gap width (the distance between two single-mode waveguides at MMI input/output) as 0.5 μm. The results show that the transmissions of TE_0 and TM_0 modes both exceed 94.4% when the MMI width is 5.2 μm. In addition, the transmission results of the two modes are still over 94% with the MMI width between 4.5 μm and 5.8 μm. Figure 7c shows the transmission versus gap width, with 5.2 μm width and 17 μm length of the MMI. The highest transmission is obtained when the gap width reaches 0.76 μm. Since the gap width determines the transverse positions of two single-mode waveguides, the highest transmission is achieved when the fundamental mode profile in two single-mode waveguides matches well with the first-order mode profile in MMI. From the above discussion, it can be concluded that this MMI has a good fabrication tolerance. Additionally, Figure 7d shows the transmission versus wavelength, where the highest value takes place at 1.6 μm. Meanwhile, the transmissions at all concerned wavelengths exceed 93.4%, demonstrating that the 1 dB bandwidth of the beam splitter includes C, L, and S telecommunication bands ranging from 1460 nm to 1625 nm.

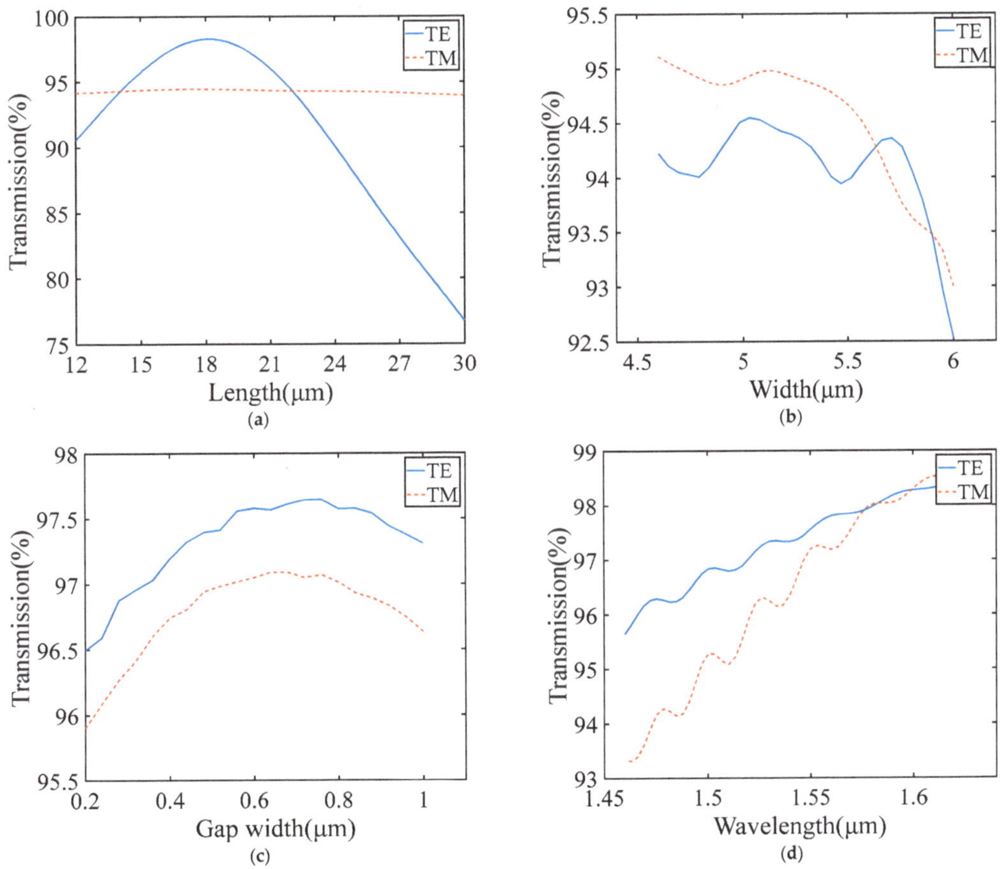

Figure 7. (**a**) The transmission of the 1 × 2 MMI with different length. (**b**) The transmission of the 1 × 2 MMI with different width. (**c**) The transmission of the 1 × 2 MMI with different gap width. (**d**) The transmission of the 1 × 2 MMI with different wavelength.

Based on the design of the 1 × 2 MMI, the 2 × 2 MMI was designed using the same method. The multimode waveguide width of the 2 × 2 MMI is preliminarily set to 6 µm. When $W_{MMI} = 6$ µm, n_0 and n_1 are 2.038 and 2.026, as determined by calculating that β_0 and β_1 are 8.260 and 8.214, respectively, the length of L_π is 68.29 µm, and the MMI is calculated as $L_{MMI} = 102.44$ µm.

As shown in Figure 8a, when the length of the MMI is 92 µm, the transmissions of both channels and modes reach a maximum, which is more than 47%. Meanwhile, the two transmissions of the outputs are approximately equal. Figure 8b shows the transmission curves of the 2 × 2 MMI with the width varying from 4 µm to 8 µm, seen on the condition of a 92 µm length of the MMI. When the MMI width is 6 µm, the transmissions of both channels and modes reach 47%. In addition, the transmissions of both channels and modes continue to exceed 45% with the MMI width between 5.85 µm and 6.15 µm. Since the MMI length heavily determines the transmission and beam-splitting ratio of the directional coupler, it impacts the interference visibility of the proposed interferometer. Therefore, we calculate the transmission, with the MMI length varying from 80 µm to 110 µm and with the width of the MMI as 6 µm.

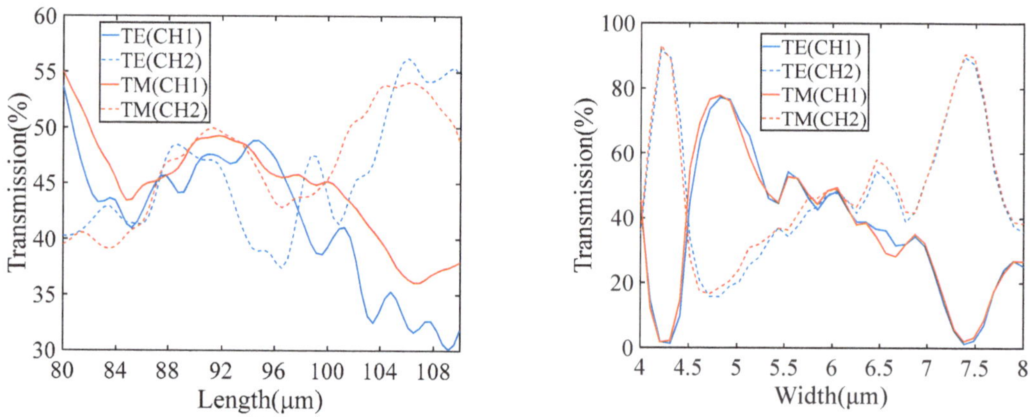

(a) (b)

Figure 8. (**a**) The transmission of the 2 × 2 MMI with different length. (**b**) The transmission of the 2 × 2 MMI with different width.

Figure 9 illustrates the definition of etching depth and sidewall angle. Figure 10a shows the 2 × 2 MMI transmission curves of TE_0 mode, with the sidewall angle varying from 60° to 90°, on the condition of a 6 µm width, 92 µm length, and 1 µm gap width. We set the upper surface width of the ridge waveguide to be invariable, so the cross-sectional area of the LN waveguide is reduced with the increase in inclination angle. The cross-sectional area varies with the inclination angle, which affects the effective refractive index; therefore, the smaller the inclination angle, the lower the transmission will be. When the angle is between 75° and 90°, the MMI has a good transmission result that exceeds 95%. With the geometry parameters unchanged, Figure 10b shows the 2 × 2 MMI transmission curves of the TE_0 mode with the etch depth varying from 0.4 to 0.6 µm. The effective refractive index is also affected by the etch depth, and so the higher the etch depth is, the lower the transmission will be. The MMI show a transmission result over 90% when the etch depth is between 0.4 and 0.5 µm. It can be concluded that this MMI has a good fabrication tolerance in terms of the inclination angle and the etch depth.

Figure 9. A schematic diagram of the sidewall angle and etching depth.

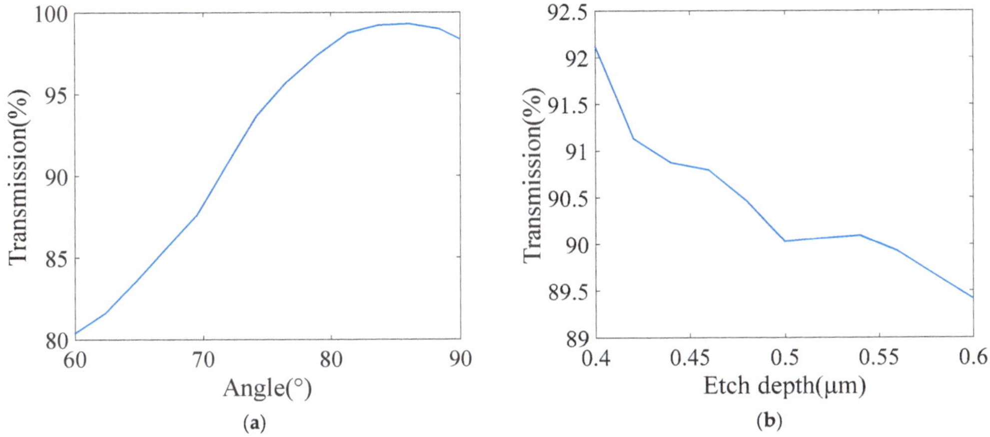

Figure 10. (**a**) The transmission of the 2 × 2 MMI with different inclination angle. (**b**) The transmission of the 2 × 2 MMI with different etch depth.

3. Performance of the Designed Polarization-Independent Interferometer

The TE_0 and TM_0 modes are injected into the input end of the beam splitter, which divides the input beam into two arms. Passing through the following mode converter, the input TE_0 and TM_0 modes become TE_0 and TE_1 modes, respectively. Then, the converted modes enter the modulation arms, one of which is imposed on a gradually increased phase shift.

For the input TE_0 mode, at 1550 nm, the transmission of the two arms at the output end was calculated, and the results are shown in Figure 11. When the two waveguides are in phase, the transmission of the output end is 49.4%. When the phase of the two waveguides is orthogonal, the transmission of the output end is 1.2% and 98.3%, respectively. The extinction ratio is calculated to be as high as 97.95%.

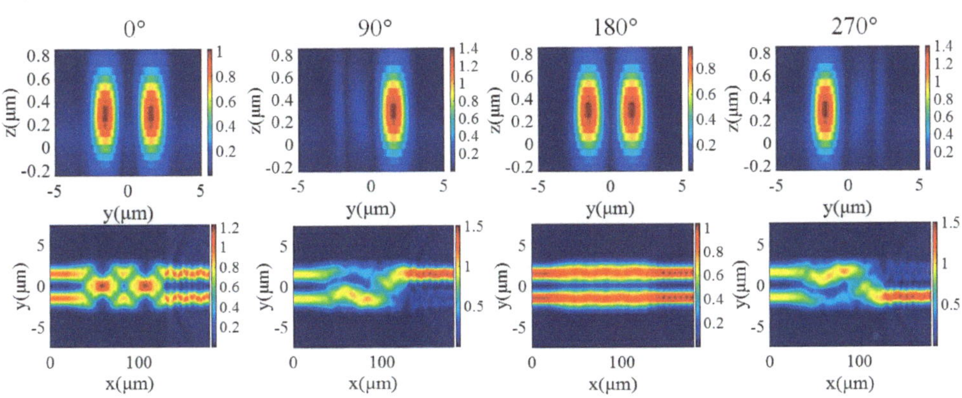

Figure 11. The output mode profile and energy distribution in the 2 × 2 MMI for TE polarization with respect to different phase shifts.

When the input is TM_0 mode at 1550 nm, the transmission of the two arms at the output end is simulated as shown in Figure 12. The transmission of the output end is 48.2% when the two waveguides are in phase. On the condition of orthogonal phase, the transmission of the output end is 0.88% and 94.95%, respectively. The extinction ratio is calculated as 98.16%, which is a little lower than that of the TE_0 mode.

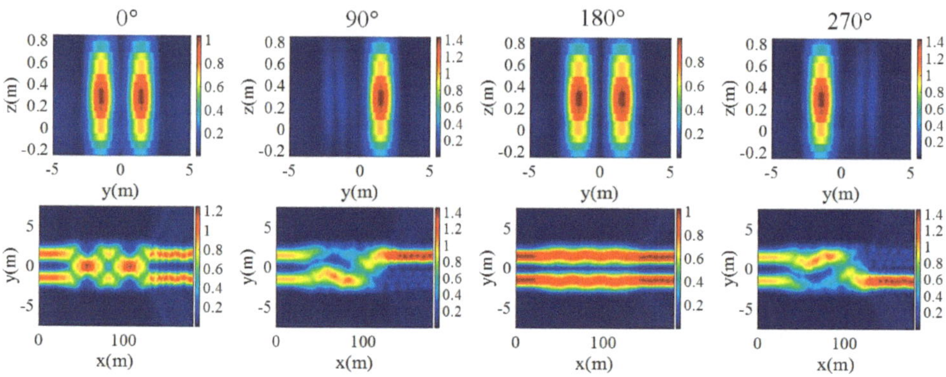

Figure 12. The output mode profile and energy distribution in the 2 × 2 MMI for TM polarization with respect to different phase shifts.

The work bandwidth for TE and TM modes was researched. The transmissions of two outputs were calculated with varying wavelengths in an orthogonal phase condition, as shown in Figure 13. Where the lowest extinction ratio takes place at 1.46 μm, the transmissions of TE are 91.16% and 7.6%, and the transmissions of TM are 89.65% and 5.1%, respectively. Therefore, the extinction ratio is more than 85%, demonstrating that the 1 dB bandwidth of the beam splitter includes C, L, and S telecommunication bands ranging from 1460 nm to 1625 nm.

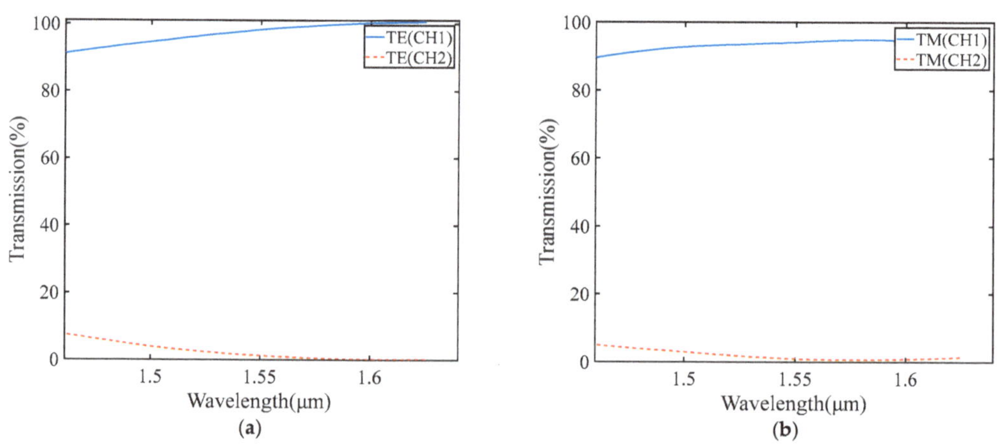

Figure 13. (**a**) The work bandwidths for TE. (**b**) The work bandwidths for the TM mode.

4. Conclusions

We propose a polarization-independent EO interferometer based on an X-CUT LN platform. This innovative design offers a promising solution for overcoming the polarization dependency typically associated with the LN interferometer. The geometric structures of the splitter and directional coupler are carefully designed within the telecommunication bands, while losses are kept at a low level. Furthermore, it exhibits significant tolerance to size variations, decreasing the pressure of fabrication. At 1550 nm, we numerically demonstrated that the extinction ratios of 97.59% and 98.16% were achieved for TE and TM modes, respectively. The proposed interferometer had prospective applications in efficient optical control, optical telecommunications, and quantum information processing owing to its simplicity in design, ease of fabrication, and ability to accommodate both polarizations.

Author Contributions: Conceptualization, J.L. (Jiali Liao); validation, W.L., J.L. (Jinrong Lan) and Z.W.; data curation, L.M.; writing—original draft preparation, L.L.; supervision, J.L. (Jiali Liao), Y.S. and Z.L.; project administration, J.L. (Jiali Liao) and Y.S. All authors have read and agreed to the published version of the manuscript.

Funding: This project was funded by the National Natural Science Foundation of China (Grant Nos. 62005207 and 61701505); 2021 Open Project Fund of Science and Technology on Electromechanical Dynamic Control Laboratory, China; the Open Research Fund of State Key Laboratory of Pulsed Power Laser Technology (Grant No. SKL 2019 KF 06); and the Natural Science Foundation of Shaanxi Province, China (Grant Nos.2019JQ-648 and 2022JM-341).

Data Availability Statement: The original contributions presented in the study are included in the article. Further inquiries can be directed to the corresponding author.

Conflicts of Interest: The authors declare no conflicts of interest.

References

1. Comellas, J.; Junyent, G. In Optical Interconnection for Datacenters: To Switch or Not to Switch. In Proceedings of the 2023 23rd International Conference on Transparent Optical Networks (ICTON), Bucharest, Romania, 2–6 July 2023; pp. 1–4.
2. Goodman, J.W.; Leonberger, F.J.; Sun-Yuan, K.; Athale, R.A. Optical interconnections for VLSI systems. *Proc. IEEE* **1984**, *72*, 850–866. [CrossRef]
3. Haurylau, M.; Chen, G.; Chen, H.; Zhang, J.; Nelson, N.A.; Albonesi, D.H.; Friedman, E.G.; Fauchet, P.M. On-Chip Optical Interconnect Roadmap: Challenges and Critical Directions. *IEEE J. Sel. Top. Quantum Electron.* **2006**, *12*, 1699–1705. [CrossRef]
4. Kahn, J.; Kazovsky, L. *Coherent Optical Communications: Fundamentals and Future Prospects, Frontiers in Optics, Rochester, New York, 2006/10/10*; Optica Publishing Group: Rochester, NY, USA, 2006; p. FThI1.
5. Miller, D.A.B. Device Requirements for Optical Interconnects to Silicon Chips. *Proc. IEEE* **2009**, *97*, 1166–1185. [CrossRef]
6. Reed, G.T.; Mashanovich, G.; Gardes, F.Y.; Thomson, D.J. Silicon optical modulators. *Nat. Photon.* **2010**, *4*, 518–526. [CrossRef]
7. Shinya, A.; Kida, K.; Sato, H.; Lu, G.W.; Yokoyama, S.; Fujikata, J. In High-Speed Optical Convolutional Neural Network Accelerator with 100 Gbaud EO-polymer/Si Hybrid Optical Modulator. In Proceedings of the 2023 Opto-Electronics and Communications Conference (OECC), Shanghai, China, 2–6 July 2023; pp. 1–4.
8. Xu, Q.; Schmidt, B.; Pradhan, S.; Lipson, M. Micrometre-scale silicon electro-optic modulator. *Nature* **2005**, *435*, 325–327. [CrossRef] [PubMed]
9. Gupta, S.; Sharma, S.; Ahmad, T.; Kaushik, A.S.; Jha, P.K.; Gupta, V.; Tomar, M. Demonstration of efficient SBN thin film based miniaturized Mach Zehnder EO modulator. *Mater. Chem. Phys.* **2021**, *262*, 124300. [CrossRef]
10. Li, C.; Zheng, W.; Dang, P.; Zheng, C.; Zhang, D. Investigation of a low-voltage polymeric 2×2 Mach-Zehnder interferometer optical switch using five-serial-coupled electro-optic microrings. *Optik* **2017**, *137*, 313–326. [CrossRef]
11. Thomas, R.; Ikonic, Z.; Kelsall, R.W. Electro-optic metal–insulator–semiconductor–insulator–metal Mach-Zehnder plasmonic modulator. *Photon. Nanostruct. Fundam. Appl.* **2012**, *10*, 183–189. [CrossRef]
12. Kieninger, C.; Kutuvantavida, Y.; Elder, D.L.; Wolf, S.; Zwickel, H.; Blaicher, M.; Kemal, J.N.; Lauermann, M.; Randel, S.; Freude, W.; et al. Ultra-high electro-optic activity demonstrated in a silicon-organic hybrid modulator. *Optica* **2018**, *5*, 739–748. [CrossRef]
13. Tang, Y.; Peters, J.D.; Bowers, J.E. Over 67 GHz bandwidth hybrid silicon electroabsorption modulator with asymmetric segmented electrode for 1.3 µm transmission. *Opt. Express* **2012**, *20*, 11529–11535. [CrossRef]
14. Bazzan, M.; Sada, C. Optical waveguides in lithium niobate: Recent developments and applications. *Appl. Phys. Rev.* **2015**, *2*, 040603. [CrossRef]
15. Li, X.; Zhao, Y.; Tao, J.; Li, R.; Liu, J.; Li, J. Linearity-Enhanced integrated lithium niobate modulator based on carrier separated asymmetric Mach−Zehnder structure. *Opt. Laser Technol.* **2024**, *176*, 110895. [CrossRef]
16. Albota, M.A.; Wong, F.N.C.; Shapiro, J.H. Polarization-independent frequency conversion for quantum optical communication. *J. Opt. Soc. Am. B* **2006**, *23*, 918–924. [CrossRef]
17. Jignesh, J.; Corcoran, B.; Schröder, J.; Lowery, A. Polarization independent optical injection locking for carrier recovery in optical communication systems. *Opt. Express* **2017**, *25*, 21216–21228. [CrossRef] [PubMed]
18. Chen, D.; Zhang, X.; Fan, K.; Wang, J.; Lu, H.; Wang, Q.; Wu, S.; Hao, R.; Li, Z.; Jin, J. Experimental demonstration of a hybrid OFDMA/NOMA scheme for multi-user underwater wireless optical communication systems. *Opt. Commun.* **2023**, *548*, 129823. [CrossRef]

19. Saxena, A. On the role of optical materials in the realization of schemes for secure quantum communication. *Mater. Today Proc.* **2023**. [CrossRef]
20. Pan, S.; Zhang, Y. Tunable and wideband microwave photonic phase shifter based on a single-sideband polarization modulator and a polarizer. *Opt. Lett.* **2012**, *37*, 4483–4485. [CrossRef] [PubMed]
21. Ganjali, M.; Esmail Hosseini, S. Effects of frequency chirping and finite extinction ratio of optical modulators in microwave photonic IFM receivers. *Opt. Commun.* **2019**, *452*, 380–386. [CrossRef]
22. Hsu, C.W.; Huang, C.F.; Tsai, W.S.; Wang, W.S. Lithium Niobate Polarization-Independent Modulator Using Integrated Polarization Splitters and Mode Converters. *J. Lightw. Technol.* **2017**, *35*, 1663–1669. [CrossRef]
23. Song, L.; Liu, W.; Guo, Z.; Li, H.; Xie, Y.; Yu, Z.; Li, H.; Shi, Y.; Dai, D. Anisotropic Thermo-Optic Mach–Zehnder Interferometer on LNOI for Polarization Handling and Multiplexing. *Laser Photon. Rev.* **2023**, *17*, 1863–8880. [CrossRef]
24. He, M.; Xu, M.; Ren, Y.; Jian, J.; Ruan, Z.; Xu, Y.; Gao, S.; Sun, S.; Wen, X.; Zhou, L.; et al. High-performance hybrid silicon and lithium niobate Mach–Zehnder modulators for 100 Gbit s−1 and beyond. *Nat. Photon.* **2019**, *13*, 359–364. [CrossRef]
25. Dai, D.; Tang, Y.; Bowers, J.E. Mode conversion in tapered submicron silicon ridge optical waveguides. *Opt. Express* **2012**, *20*, 13425–13439. [CrossRef] [PubMed]
26. Daoxin, D.; Sailing, H.; Hon-Ki, T. Bilevel mode converter between a silicon nanowire waveguide and a larger waveguide. *J. Lightw. Technol.* **2006**, *24*, 2428–2433. [CrossRef]
27. Zhang, J.; Qiu, P.; He, R.; Song, X.; Dai, Z.; Liu, Y.; Pan, D.; Yang, J.; Guo, K. Compact mode converters in thin-film lithium niobate integrated platforms. *Opt. Lett.* **2024**, *49*, 2958–2961. [CrossRef] [PubMed]
28. Calò, G.; Bellanca, G.; Fuschini, F.; Barbiroli, M.; Bertozzi, D.; Tralli, V.; Petruzzelli, V. 4 × 4 Integrated Switches Based on On-Chip Wireless Connection through Optical Phased Arrays Photonics. *Photonics* **2023**, *10*, 367. [CrossRef]
29. Liu, X.; Zhao, Y.; Zhu, Z.; Liu, H.; Gan, F. Particle Swarm Optimized Compact, Low Loss 3-dB Power Splitter Enabled by Silicon Columns in Silicon-on-Insulator Photonics. *Photonics* **2023**, *10*, 419. [CrossRef]

Disclaimer/Publisher's Note: The statements, opinions and data contained in all publications are solely those of the individual author(s) and contributor(s) and not of MDPI and/or the editor(s). MDPI and/or the editor(s) disclaim responsibility for any injury to people or property resulting from any ideas, methods, instructions or products referred to in the content.

micromachines

Article

Preparation of High-Performance Transparent Al_2O_3 Dielectric Films via Self-Exothermic Reaction Based on Solution Method and Applications

Xuecong Fang [1], Honglong Ning [1], Zihan Zhang [1], Rihui Yao [1,*], Yucheng Huang [1], Yonglin Yang [1], Weixin Cheng [1], Shaojie Jin [1], Dongxiang Luo [2,3,4,*] and Junbiao Peng [1]

1 Guangdong Basic Research Center of Excellence for Energy & Information Polymer Materials, State Key Laboratory of Luminescent Materials and Devices, School of Materials Sciences and Engineering, South China University of Technology, Guangzhou 510640, China; 202030270402@mail.scut.edu.cn (X.F.); ninghl@scut.edu.cn (H.N.); 201930160189@mail.scut.edu.cn (Z.Z.); yucheng_h@163.com (Y.H.); 202420119057@mail.scut.edu.cn (Y.Y.); ms_weixin_cheng@mail.scut.edu.cn (W.C.); 18770602788@163.com (S.J.); psjbpeng@scut.edu.cn (J.P.)
2 Huangpu Hydrogen Innovation Center/Guangzhou Key Laboratory for Clean Energy and Materials, School of Chemistry and Chemical Engineering, Guangzhou University, Guangzhou 510006, China
3 Guangdong Provincial Key Laboratory of Optical Information Materials and Technology, South China Academy of Advanced Optoelectronics, South China Normal University, Guangzhou 510006, China
4 Key Laboratory of Optoelectronic Devices and Systems of Ministry of Education and Guangdong Province, College of Physics and Optoelectronic Engineering, Shenzhen University, Shenzhen 518060, China
* Correspondence: yaorihui@scut.edu.cn (R.Y.); luodx@gzhu.edu.cn (D.L.)

Abstract: As the competition intensifies in enhancing the integration and performance of integrated circuits, in accordance with the famous Moore's Law, higher performance and smaller size requirements are imposed on the dielectric layers in electronic devices. Compared to vacuum methods, the production cost of preparing dielectric layers via solution methods is lower, and the preparation cycle is shorter. This paper utilizes a low-temperature self-exothermic reaction based on the solution method to prepare high-performance Al_2O_3 dielectric thin films that are compatible with flexible substrates. In this paper, we first established two non-self-exothermic systems: one with pure aluminum nitrate and one with pure aluminum acetylacetonate. Additionally, we set up one self-exothermic system where aluminum nitrate and aluminum acetylacetonate were mixed in a 1:1 ratio. Tests revealed that the leakage current density and dielectric constant of the self-exothermic system devices were significantly optimized compared to the two non-self-exothermic system devices, indicating that the self-exothermic reaction can effectively improve the quality of the dielectric film. This paper further established two self-exothermic systems with aluminum nitrate and aluminum acetylacetonate mixed in 2:1 and 1:2 ratios, respectively, for comparison. The results indicate that as the proportion of aluminum nitrate increases, the overall dielectric performance of the devices improves. The best overall performance occurs when aluminum nitrate and aluminum acetylacetonate are mixed in a ratio of 2:1: The film surface is smooth without cracks; the surface roughness is 0.747 ± 0.045 nm; the visible light transmittance reaches up to 98%; on the basis of this film, MIM devices were fabricated, with tested leakage current density as low as 1.08×10^{-8} A/cm^2 @1 MV and a relative dielectric constant as high as 8.61 ± 0.06, demonstrating excellent electrical performance.

Keywords: low temperature; self-exothermic reaction; MIM device; transparent; low leakage current; high dielectric constant

check for updates

Citation: Fang, X.; Ning, H.; Zhang, Z.; Yao, R.; Huang, Y.; Yang, Y.; Cheng, W.; Jin, S.; Luo, D.; Peng, J. Preparation of High-Performance Transparent Al_2O_3 Dielectric Films via Self-Exothermic Reaction Based on Solution Method and Applications. *Micromachines* **2024**, *15*, 1140. https://doi.org/10.3390/mi15091140

Academic Editor: Chengyuan Dong

Received: 21 August 2024
Revised: 9 September 2024
Accepted: 9 September 2024
Published: 11 September 2024

Copyright: © 2024 by the authors. Licensee MDPI, Basel, Switzerland. This article is an open access article distributed under the terms and conditions of the Creative Commons Attribution (CC BY) license (https://creativecommons.org/licenses/by/4.0/).

1. Introduction

With the advanced development of integrated circuits, there is a growing demand for thinner dielectric layers in electronic components [1–3]. However, when the thickness of the amorphous silicon dielectric layer is reduced to below 3 nm, the gate leakage current caused by direct tunneling can reach up to 1 A/cm^2 [4]. To reduce leakage current and increase the thickness of the dielectric layer, various high-k dielectric materials have been used, including Al_2O_3, HfO_2, ZnO, and ZrO_2 [5–9]. They also offer advantages such as high mobility, high transparency, and suitability for flexible substrates [10–12].

The preparation method of high-k gate dielectric layers is the classical vacuum method [13–16], resulting in excellent film uniformity and high quality. However, these equipment options are characterized by high costs and lengthy experimental durations. Therefore, in order to achieve low-cost, rapid, and high-quality preparation of dielectric layers, the solution method has attracted widespread attention and research [17–20]. However, in the process of film preparation using solution methods, in order to achieve the removal of organic ligands and densification of the film, the annealing temperature is typically higher than 400 °C or even higher [21,22]. This limits the preparation of dielectric films on flexible substrates.

Recently, many new low-temperature solution methods for preparing dielectric layers have been studied. Wangying Xu and her team prepared Al_2O_3 dielectric films using a 'aqueous route' with annealing at 250 °C [23]. The leakage current density is 2.9×10^{-7} A/cm^2@1 MV/cm, the dielectric constant is 8.6, and the breakdown field strength is greater than 2.5 MV/cm. Jeong-Wan Jo employed a high-density UV (DUV) treatment-assisted exothermic process using low-pressure mercury lamps to prepare Al_2O_3/ZrO_2 dielectric layers at 180 °C [24]. The leakage current density was reduced to 4.68×10^{-9} A/cm^2@1 MV/cm. Sumei Wang used infrared irradiation for 40 min to induce the low-temperature decomposition of $AlCl_3$ precursor solution into Al_2O_3 films, achieving a high dielectric constant [7,8] and a low leakage current (3.5×10^{-8} A/cm^2@1 MV/cm) [25]. A breakdown field strength greater than 3 MV/cm was obtained. These low-temperature solution methods share a common characteristic, which is the utilization of photo-irradiation or other methods to assist the exothermic process based on exothermic reaction. A drawback is the need for constant adjustment of the power and wavelength of auxiliary equipment to achieve optimal settings.

In this paper, we utilize a low-temperature self-exothermic reaction based on the solution method to prepare high-performance Al_2O_3 dielectric thin films. In the exothermic reaction, metal nitrates act as strong oxidants, while acetylacetonates serve as strong reductants. The mixed sample undergoes a spontaneous and vigorous redox reaction upon reaching a relatively low annealing temperature, generating a substantial amount of heat to compensate for the energy required for the formation of M-O-M bonds, thereby producing high-performance dielectric films. This exothermic process requires only a constant-temperature heating device and no additional auxiliary equipment, and the low annealing temperature allows compatibility with flexible substrates and reduces energy consumption. This low-temperature solution method offers advantages such as low cost, adjustable composition, safety, and the ability to produce large-area films [26–28]. Here, this paper proposes a self-exothermic reaction, distinct from traditional exothermic reactions, where the process only requires the mixing of metal precursors containing fuel ligands and oxidant ligands without the need for additional fuel; this simple precursor solution composition is easy to control and replicate, facilitating the preparation of stable dielectric layers and MIM (Metal–Insulator–Metal) devices.

2. Experiment

2.1. Synthesis of Precursor Solution

All reagents, including aluminum nitrate hydrate ($Al(NO_3)_3{\cdot}9H_2O$, AlNO), aluminum triacetylacetone ($C_{15}H_{21}O_6Al$, AlAC) and N,N-Dimethylformamide (C_3H_7NO, DMF), were purchased from Merck and required no further purification.

The preparation of the self-exothermic precursor solution is as follows: The total concentration is controlled at 0.2 M, which can be calculated as follows:

$$c\ (AlNO) + c\ (AlAC) = 0.2\ M \tag{1}$$

Five samples of precursor solutions were prepared by mixing AlNO and AlAC in different ratios, as shown in Table 1. The solvent chosen is DMF. It was stirred magnetically for 24 h, followed by aging for 4 h at rest. Subsequently, the precursor solution was filtered using a 0.22 μm organic-phase syringe filter to obtain the solution for spin-coating.

Table 1. Proportion of precursor system components and naming of precursor systems.

Proportion of Precursor System Components	c(AlNO)/mol	c(AlAC)/mol	Naming
Pure AlNO	0.2000	0	S1
Pure AlAC	0	0.2000	S2
AlNO:AlAC = 2:1	0.1333	0.0667	S3
AlNO:AlAC = 1:1	0.1000	0.1000	S4
AlNO:AlAC = 1:2	0.0667	0.1333	S5

In this paper, the five samples of precursor systems will be simplified, with their names shown in Table 1.

2.2. Dielectric Film Fabrication

ITO glass substrates were subjected to ultrasonic cleaning in a sequence of deionized water/isopropanol/deionized water/isopropanol, then dried at 80 °C for 12 h. Prior to spin-coating, the substrates were UV-treated for 20 min. Subsequently, 40 μL of the precursor solution was spin-coated on the substrates at 500 r/min for 5 s and 5000 r/min for 30 s. Finally, dielectric films were pre-annealed at 80 °C for 10 min and post-annealed at 185 °C for 1 h.

The weight and heat changes of the precursor solution were measured using a Thermogravimetric Analyzer (TG, DZ-TGA101, Nanjing, China) and Differential Scanning Calorimetry (DSC, DZ-DSC300C, Nanjing, China). The surface tension of the precursor solution was measured using Attension Theta Lite (TL200, BiolinScientific, Gothenburg, Sweden). The surface morphology of the dielectric films was observed using Laser Scanning Confocal Microscopy (LSCM, OLS50-CB, Tokyo, Japan) and Atomic Force Microscopy (AFM, BY3000, Nano Instruments, Guangzhou, China). The thickness, density, and roughness of the films were fitted using an X-ray Reflectometer (XRR, PANalytical Empyrean, Almelo, The Netherlands). The optical properties were studied using an Ultraviolet-Visible Spectrophotometer (UV-Vis, UV-3600 Shimadzu, Kyoto, Japan). The phase structure of the films was analyzed using an X-ray Diffractometer (XRD, PANalytical Empyrean, Almelo, The Netherlands), and the functional groups were characterized using Fourier Transform Infrared Spectroscopy (FTIR, ATR Accessory, Nexus, Madison, WI, USA).

2.3. MIM Device Fabrication

The electrical properties of dielectric films are typically characterized by preparing MIM devices [29]. MIM devices were fabricated with Al_2O_3 as the dielectric layer, as shown in Figure 1. Al electrodes were thermally deposited on the surface of the dielectric film, with an electrode area of 1.256×10^{-3} cm^2. The current–voltage (I–V), capacitance–voltage (C-V), and capacitance–frequency (C-F) characteristics of MIM devices were measured using the semiconductor parameter analyzer (Primarius FS-Pro, Shanghai, China). In order to ensure the authenticity and universality of the experimental data, three test samples were prepared for each component, and two different test points were characterized for each sample. If the test results obtained are close, it indicates good uniformity of the film.

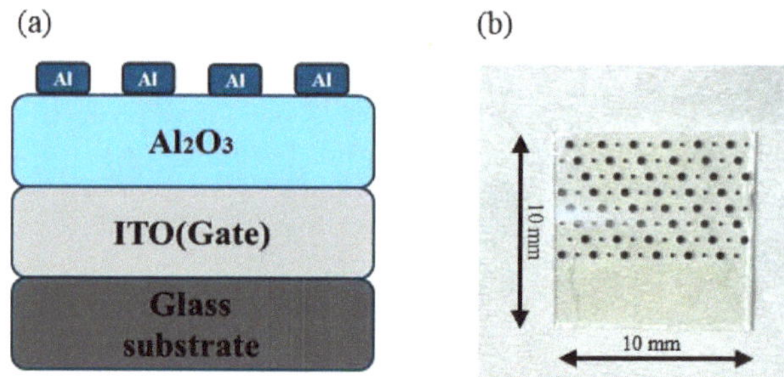

Figure 1. MIM device: (**a**) schematic; (**b**) photo.

3. Result and Discussion

3.1. Surface Tension Tests Results

The surface tension tests of the five precursor solutions are shown in Figure 2. The results show that all five samples of precursor solutions have relatively average surface tension. The surface tensions of samples S1, S2, S3, S4 and S5 studied in this paper are 35.77 mN/m, 36.58 mN/m, 35.37 mN/m, 35.67 mN/m, and 36.76 mN/m, aligning with spin-coating thin films.

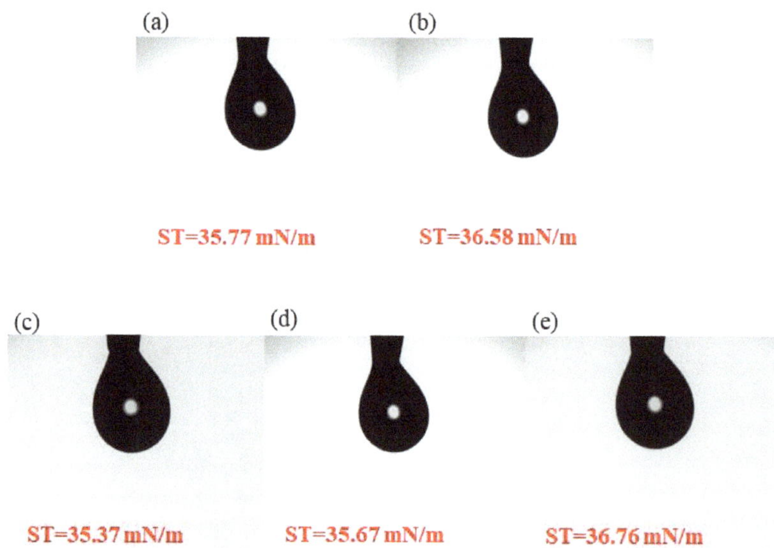

Figure 2. Surface tension: (**a**) S1; (**b**) S2; (**c**) S3; (**d**) S4; (**e**) S5.

3.2. TG-DSC Test Analysis

The TG-DSC test results are shown in Figure 3. From the results, it is evident that three groups of self-exothermic precursor samples exhibit a distinctly different thermal behavior compared to the other two non-self-exothermic precursor samples. S3, S4, and S5 exhibit intense exothermic behavior near 160~180 °C. In contrast, S1 and S2 show no significant exothermic activity throughout the entire temperature range (30~500 °C).

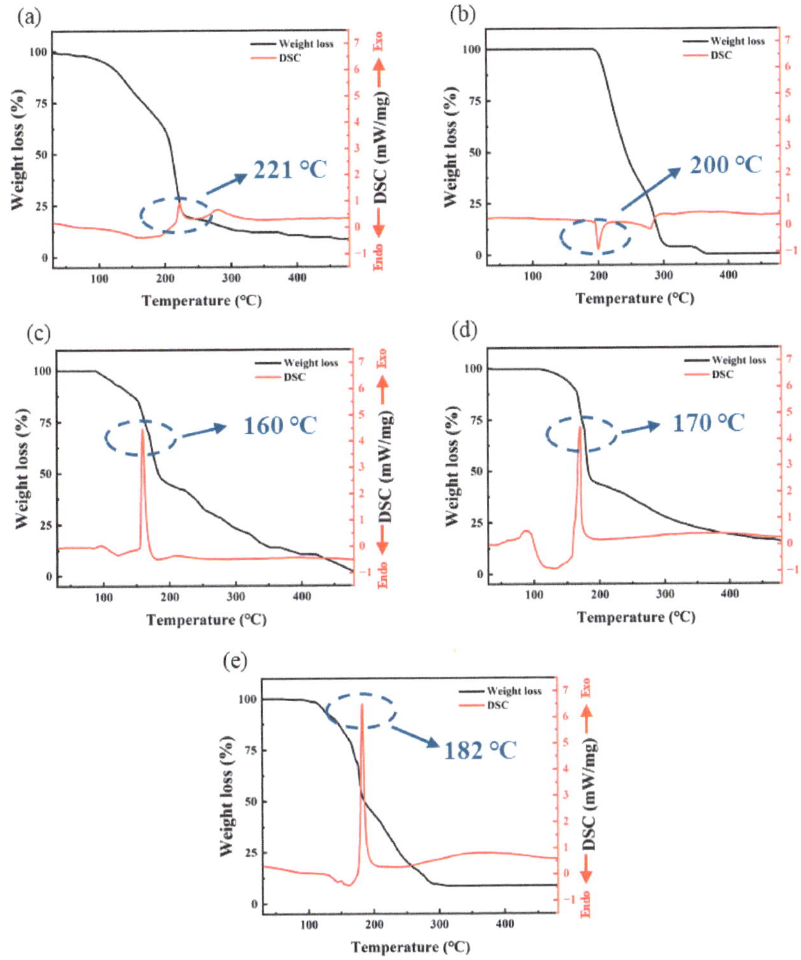

Figure 3. The TG-DSC test curve: (**a**) S1; (**b**) S2; (**c**) S3; (**d**) S4; (**e**) S5.

The decomposition temperature of $Al(NO_3)_3$ is around 200 °C, resulting in the formation of Al_2O_3 and the release of a large amount of gasses such as NO, NO_2, and O_2 [30]. From Figure 3a, it can be observed that S1 experiences a sharp decrease in mass around 220 °C, corresponding to the production of a large amount of gas during the decomposition reaction. Therefore, the exothermic behavior of S1 around 221 °C is likely due to the decomposition reaction. S2 exhibits a weak endothermic peak around 200 °C, which is due to the melting and sublimation of AlAC at this temperature, with a significant mass loss resulting from the evaporation of a substantial amount of material.

However, three samples of different self-exothermic precursors exhibited a significant exothermic peak at 160~180 °C. S3, S4 and S5, respectively, exhibited sharp exothermic peaks around 160 °C, 170 °C, and 182 °C, with exothermic intensities of 4.46 mW/mg, 4.48 mW/mg and 6.53 mW/mg, respectively. Each exothermic peak corresponds to a significant mass loss, which is due to the generation of large amounts of gas from redox reactions. The temperature corresponding to the exothermic peak increases with an increase in the proportion of AlAC; at the same time, the exothermic intensity also increases. Additionally, S4 exhibited a broad endothermic peak around 138 °C. This is due to the residual DMF solvent in the sample boiling and absorbing heat at this temperature.

From the DSC data of S1 and S2, it can be observed that neither S1 nor S2 exhibits significant exothermic behavior below 200 °C. This indicates that the exothermic behavior observed in S3, S4, and S5 in the range of 160~185 °C is not due to a single substance of AlNO or AlAC but rather results from the redox reaction between AlNO and AlAC: at this temperature, the nitrate ions in AlNO, along with a small amount of oxygen generated from the decomposition of nitrate ions, act as strong oxidants. They undergo vigorous redox reactions with AlAC, acting as a reducing agent, producing a large amount of heat, resulting in the formation of Al_2O_3 and a substantial quantity of gas. Significant mass loss observed in the TGA test results is attributable to this reaction. The heat provided by this rapid and intense exothermic reaction may promote the formation of M-O-M bonds and the removal of organic ligands, offering the potential for the preparation of high-performance dielectric layers at low temperatures.

In Figure 4, DSC data for other aluminum salts are provided for comparison. As seen in Figure 4a, a prominent endothermic peak appears around 130 °C, which is attributed to the removal of H_2O in sample $Al(NO_3)_3{\cdot}8H_2O$ [30]. In contrast, no similar endothermic peak appears around 130 °C in Figure 3a, as the DSC samples in this article were first dissolved in DMF and then dried to form solid samples, so the H_2O was already dissolved in the DMF. However, none of the aluminum salts in Figure 4 exhibit significant exothermic peaks, which further indicates that the prominent exothermic peaks observed in Figure 3c–e are not caused by a single aluminum salt. Instead, they are due to the redox reactions between AlNO and AlAC, which release a substantial amount of heat.

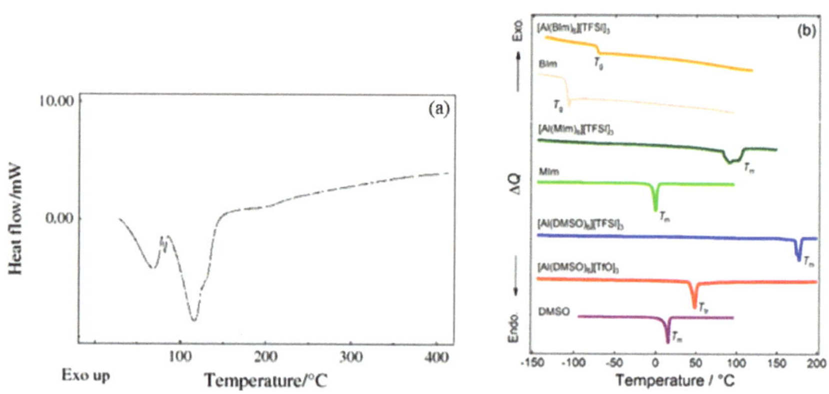

Figure 4. Comparison of DSC curves for different aluminum salts: (**a**) DSC curve of $Al(NO_3)_3{\cdot}8H_2O$ [31]; (**b**) DSC traces of $[Al(L)_6]X_3$ (L = DMSO, MIm, or BIm; X = TFSI or TfO) with the pure ligands as references. T_m, T_{tr}, and T_g denote melting points, solid-solid phase transition temperatures, and glass transition temperatures, respectively [32].

3.3. *Surface Morphology Analysis*

As shown in Figure 5, the surface morphology of different film samples was observed using LSCM. From Figure 5a,c,d, the film surfaces are smooth and uniform, with no visible cracks or black particles. In contrast, Figure 5b,e clearly shows a significant distribution of black particles on the film surfaces, leading to an uneven texture. This may be due to the low solubility of AlAC in the DMF solvent: as the temperature increases, AlAC precipitates out before it has a chance to react, resulting in the formation of crystalline particles on the film surface.

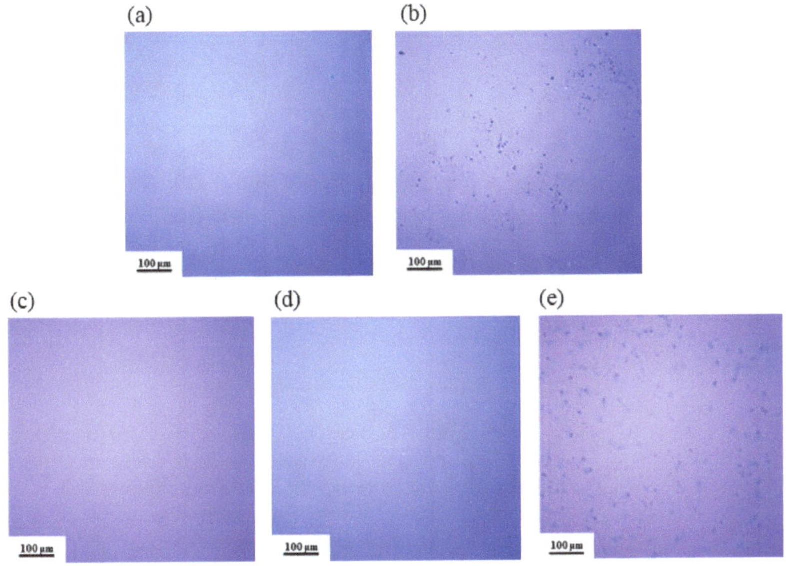

Figure 5. LSCM images of Al_2O_3 films: (**a**) S1; (**b**) S2; (**c**) S3; (**d**) S4; (**e**) S5.

Surface morphology of the thin films was scanned using AFM, as show in Figure 6, and surface roughness was calculated, as shown in Figure 7. The roughness of all five samples of dielectric films is less than 1 nm, indicating good uniformity of the dielectric films prepared by the solution method, providing potential for the fabrication of high-performance devices.

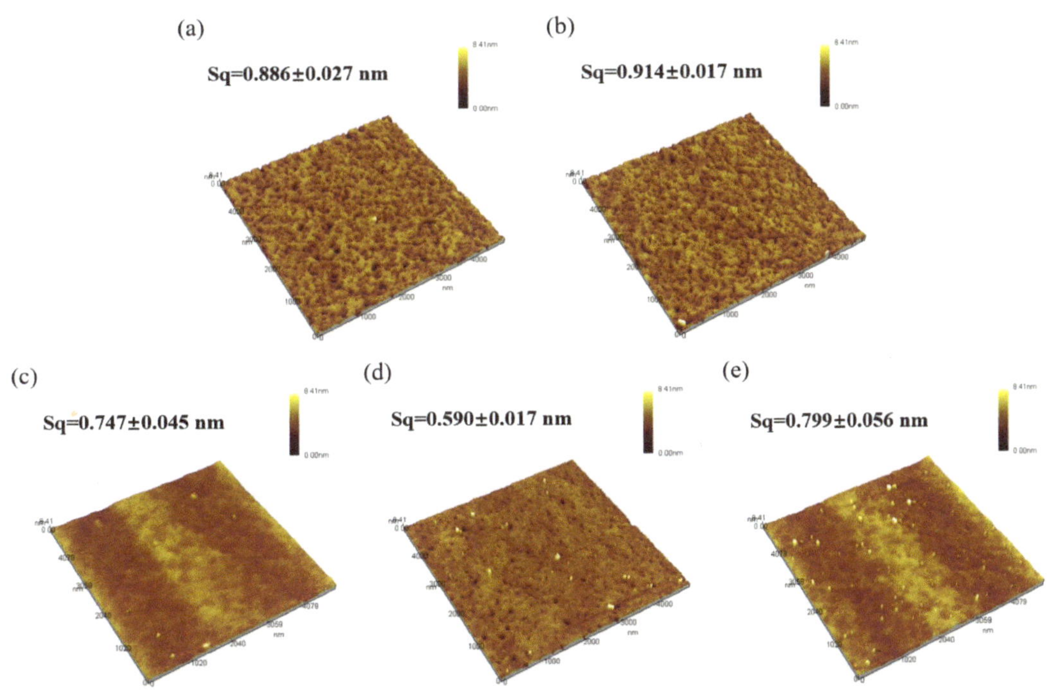

Figure 6. AFM morphology of Al_2O_3 films: (**a**) S1; (**b**) S2; (**c**) S3; (**d**) S4; (**e**) S5.

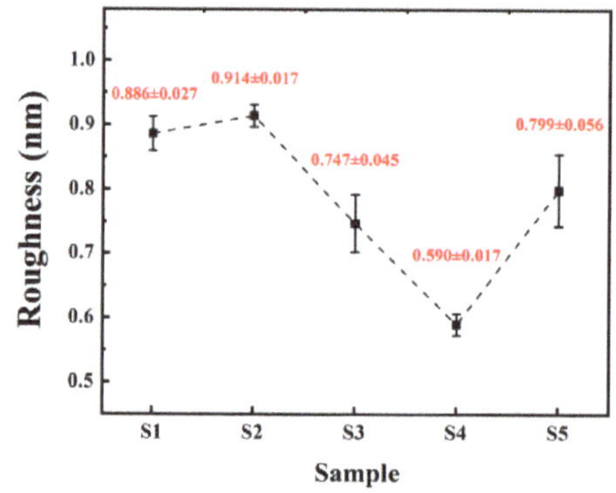

Figure 7. Surface roughness of different samples.

From Figure 6, it can be observed that both samples of non-self-exothermic system dielectric film surfaces exhibit relatively high roughness and numerous surface pores. This is due to the insufficient annealing temperature, resulting in a lower degree of film densification and, thus, the formation of more pores. In contrast, the surface roughness of the dielectric films in the three self-exothermic system samples is lower than that of the non-self-exothermic system samples. This indicates that the significant heat generated during annealing in the self-exothermic system contributes to an increased level of film densification. The surface roughness of samples S3 and S5 reached 0.747 ± 0.045 nm and 0.799 ± 0.056 nm, respectively. Meanwhile, the surface roughness of sample of S4 reached the lowest at 0.590 ± 0.017 nm, demonstrating the best performance. From Figure 6e, it can also be observed that there is a significant amount of particle precipitation on the surface of the S5 dielectric film. This is consistent with the results of LSCM testing, indicating that a higher content of AlAC leads to easier precipitation during annealing.

3.4. Thin Film Thickness Measurement

The insulation performance is closely related to the thickness, roughness, and density of the dielectric films [33]. XRR was used to fit the thickness and density, as shown in Figure 8. It is found that the thickness of samples of S1, S2, S3, S4 and S5 is 8.772 nm, 9.322 nm, 10.328 nm, 9.325 nm, and 7.872 nm.

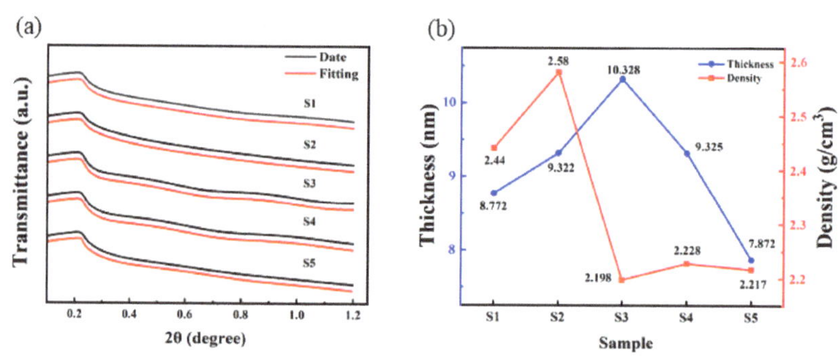

Figure 8. Al_2O_3 films: (**a**) XRR date and fitting curves; (**b**) thickness and density fitting results.

3.5. Transmittance Test

The transmittance of the dielectric films was measured using UV-vis, and the test results are shown in Figure 9. Within the visible light range (400–760 nm), all five samples of films exhibited transmittance rates of over 98%, demonstrating excellent optical performance.

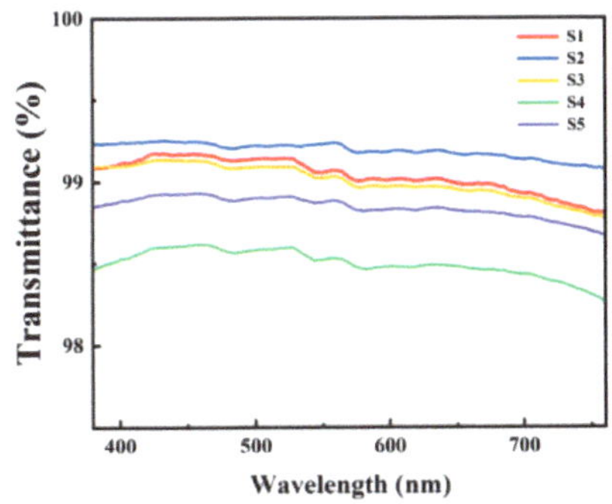

Figure 9. UV-Vis spectrum of different Al_2O_3 films.

3.6. Leakage Current Density Analysis

Figure 10 displays the I–V testing curves of five sets of dielectric thin films. Within the tested range, the leakage current density of S3 and S4 is lower than that of S1 and S2, with S3 exhibiting the lowest leakage current density in the 0 to 5 MV/cm range. This indicates that a self-exothermic reaction contributes to the formation of M-O-M bonds within the film and the reduction in defect state density. Conversely, at higher electric fields, the leakage current density of S5 approaches that of S1 and S2, which may be attributed to the higher surface roughness of the S5 film, leading to increased defects in the contact area between the electrode and the film, thereby increasing the leakage current. Furthermore, the zero points of the IV curves for S3 are primarily located around 0 MV/cm, whereas the zero points for S1, S2, and S4 are shifted to around −2 to −1 MV/cm. This also indicates that a self-exothermic reaction helps to reduce internal defects within the film, thereby suppressing zero-point drift.

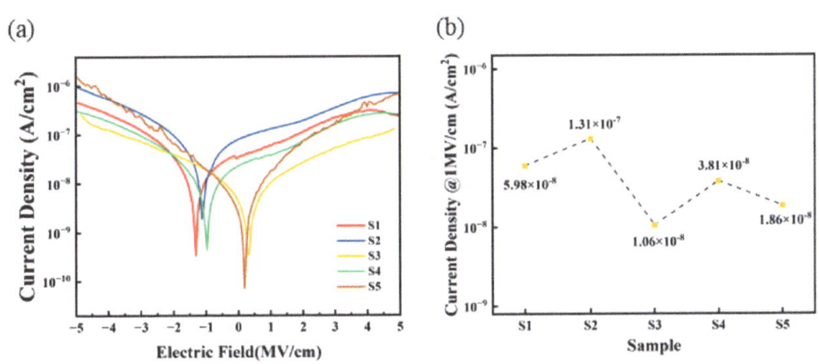

Figure 10. Al_2O_3 dielectric films: (**a**) leakage current density curves; (**b**) summary of leakage current density @1 MV/cm.

When an electric field of 1 MV/cm is applied, the leakage current densities of the three exothermic system (S3, S4, and S5) devices are 1.06×10^{-8} A/cm^2, 3.81×10^{-8} A/cm^2, and 1.86×10^{-8} A/cm^2, respectively.

3.7. Capacitance Density Analysis

Capacitance density (C_i) is one of the important parameters of the dielectric layer, reflecting the capacity of storing electrical charge per unit area of the film. The results of the C-V tests are shown in Figure 11. From the results, within the test range, S3 and S4 exhibit good capacitance–voltage stability and higher C_i. However, the C_i of the two non-exothermic system devices is smaller, indicating that the significant heat released by the exothermic process benefits the formation of M-O-M bonds within the dielectric film. During the experiment, the C_i of S5 was so low that it exceeded the instrument's range, resulting in a value close to zero. Combined with the unstable C_i observed in S2 and the results of surface roughness tests, we can conclude that even with exothermic reactions occurring, the excessive surface roughness of the film significantly affects the performance and stability of the devices.

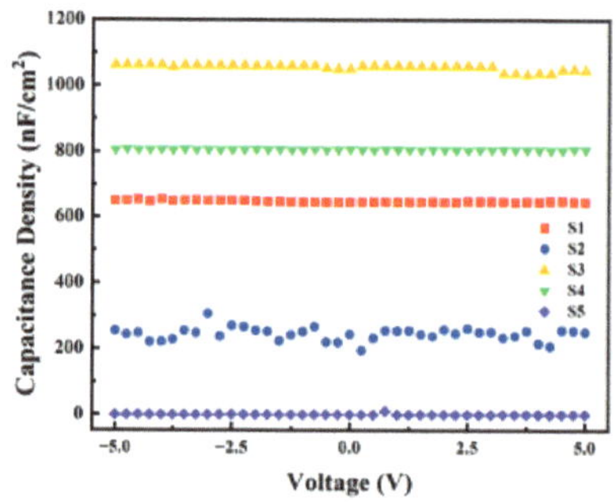

Figure 11. Capacitance density test curves of different samples.

The relative dielectric constant of dielectric films is further investigated. C-F tests were conducted on MIM devices. The relative dielectric constant can be calculated based on the C-F test results and the thickness of dielectric films, which can be calculated as follows:

$$C = \varepsilon_0 k \frac{S}{d} \quad (2)$$

where ε_0 is the vacuum permittivity (8.85×10^{-12} F/m), d represents the thickness of films, and s denotes the area of Al electrode (1.256×10^{-3} cm^2). The relative dielectric constants @ 1 kHz of dielectric films are summarized in Figure 12. The relative dielectric constants of the three exothermic system films are all greater than those of the other two non-exothermic system films, which once again demonstrates the promoting effect of exothermic reactions. S3 exhibits the highest relative dielectric constant (8.61 ± 0.06) and the smallest standard deviation, indicating excellent dielectric performance and uniformity of the film.

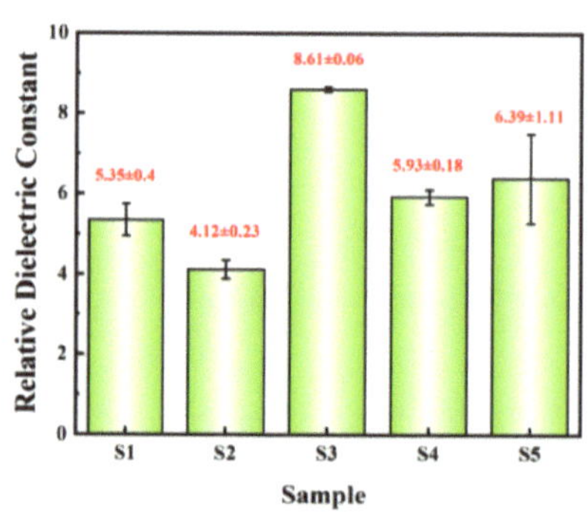

Figure 12. Summary of relative dielectric constant @1 kHz.

Finally, I have listed a comparison of the key performance characteristics between the films discussed in this article and those produced using the most advanced methods currently available in Table 2. From the table, the Al_2O_3 films in this study exhibit superior performance in terms of leakage current density and dielectric constant compared to those prepared using auxiliary equipment. At the same time, the Al_2O_3 films in this article show performance in leakage current density that is comparable to films prepared via vacuum deposition methods and even surpasses the performance of some individual films. Therefore, the low-temperature self-exothermic reaction method for preparing dielectric films based on solution processing presented in this article is not only easy to operate and requires simple equipment but also demonstrates excellent performance, showing great potential. However, this method still faces several issues and challenges. Firstly, since the method is based on solution processing, the evaporation of the solvent during the annealing process can result in the formation of numerous pores of varying sizes within the film. This may lead to reduced film density and adhesion to substrates and increase leakage current density. Secondly, the organic solvents used in this study cause the formation of more pores on the film's surface during annealing, which increases the surface roughness and thus affects the electrical properties of the film.

Table 2. A thorough quantitative comparison of leakage current density and dielectric constant against state-of-the-art methods.

Dielectric Material	Preparation Method	Leakage Current Density (A/cm^2)	Relative Dielectric Constant	Refs.
Al_2O_3	self-exothermic reaction	1.06×10^{-8} @1 MV/cm	8.61 @1000 Hz	this article
Al_2O_3	DUV assisted exothermic process	2.9×10^{-7} @1 MV/cm	8.6 @1000 Hz	[23]
Al_2O_3	aqueous route	4.68×10^{-9} @1 MV/cm	8.6 @100 Hz	[24]
Al_2O_3	infrared irradiation	3.5×10^{-8} @1 MV/cm	7.6 @1000 Hz	[25]
$Al_2O_3/TiO_2/Al_2O_3$ (6 nm/40 nm/6 nm)	PLD	2.03×10^{-8} @3 V		[34]
$Al_2O_3/HfO_2/Al_2O_3$	ALD	10^{-8}~10^{-9} @1 MV/cm	20.7	[35]
Al_2O_3	PEALD	10^{-9} @1 MV/cm	9.3 @100 kHz	[36]

4. Conclusions

This paper employs a low-temperature self-exothermic reaction based on the solution method to fabricate high-performance transparent Al_2O_3 dielectric films at an annealing temperature of 185 °C. The experiments demonstrate that mixed systems of AlNO and AlAC in varying proportions undergo vigorous redox reactions and release significant heat at around 160~180 °C. During the annealing process, a higher content of AlAC results in more pronounced surface precipitation and higher surface roughness of the film. Therefore, based on comprehensive analysis, S3 exhibits the best performance: the AFM test results indicate a surface roughness of 0.747 ± 0.045 nm; the transmittance within the visible light range reaches up to 98%; the device leakage current density is as low as 1.06×10^{-8} A/cm^2@1 MV/cm, with a high relative dielectric constant of 8.61 ± 0.06, demonstrating excellent electrical characteristics and stability. In summary, the alumina films prepared using the exothermic solution method reported in this paper exhibit lower annealing temperatures and surface roughness, higher transparency, lower leakage current density, and higher relative dielectric constant. These characteristics offer great promise for compatibility with flexible substrates and the fabrication of high-performance transparent electronic devices.

Author Contributions: Writing—original draft preparation, X.F.; writing—review and editing, R.Y. and H.N.; methodology, Z.Z.; conceptualization, H.N.; investigation, W.C. and S.J.; validation, J.P. and Y.H.; formal analysis, Y.Y.; supervision, D.L. All authors have read and agreed to the published version of the manuscript.

Funding: This work was supported by the National Key R&D Program of China (No. 2021YFB3600604), the National Natural Science Foundation of China (Grant No. 62174057, 62375057, 62074060 and 22090024), the Guangdong Natural Science Foundation (No. 2024A1515012216 and 2023A1515011026), the Educational Commission of Guangdong Province (Grant Nos. 2022ZDZX1002), and the Key R&D Plan of Guangdong Province (2022B0303010001), State Key Lab of Luminescent Materials and

Devices (Skllmd-2024-05), Southwest Institute of Technology and Engineering Cooperation Fund (HD-HDW59A020301), the open research fund od Songshan Lake Materials Laboratory (2022SLABFN05), Science and Technology Program of Guangdong (Grant No. 2024A0505040026), Guangdong Basic and Applied Basic Research Foundation (Grant No. 2022A1515140064, 2023B1515120046 and 2024A1515012019), Science and Technology Program of Guangzhou (Grant No. 2023A03J0024), Research Project of Guangzhou University (Grant No. ZH2023006), the Guangdong Provincial Key Laboratory of Optical Information Materials and Technology (2023B1212060065), and the Key Laboratory of Optoelectronic Devices and Systems of Ministry of Education and Guangdong Province.

Data Availability Statement: The data are contained within the article.

Acknowledgments: We thank all the participants, institutions, editors, and reviewers for enabling us to conduct this research.

Conflicts of Interest: The authors declare no conflicts of interest.

References

1. Yi, Z.; Lyu, B.; Gao, D.; Hao, J.; Liu, Z.; Ning, Z.; Jiang, B. Preparation Strategy and Composition Design of Polymer-Based Layered Composites for Improving Energy Storage Performances. *J. Energy Storage* **2024**, *92*, 111995. [CrossRef]
2. Huan, T.D.; Boggs, S.; Teyssedre, G.; Laurent, C.; Cakmak, M.; Kumar, S.; Ramprasad, R. Advanced Polymeric Dielectrics for High Energy Density Applications. *Prog. Mater. Sci.* **2016**, *83*, 236–269. [CrossRef]
3. Yao, Z.; Song, Z.; Hao, H.; Yu, Z.; Cao, M.; Zhang, S.; Lanagan, M.T.; Liu, H. Homogeneous/Inhomogeneous-Structured Dielectrics and Their Energy-Storage Performances. *Adv. Mater.* **2017**, *29*, 1601727. [CrossRef] [PubMed]
4. Mukherjee, A.; Nagaria, R.K. A Comparative Study of Various High-K Dielectric Materials as Gate Oxide Layer on TG FinFET. In Proceedings of the 2021 2nd Global Conference for Advancement in Technology (GCAT), Bangkok, Thailand, 1–3 October 2021; pp. 1–4. [CrossRef]
5. Han, D.; Geng, Y.; Cai, J.; Wang, W.; Wang, L.; Tian, Y.; Wang, Y.; Liu, L.; Zhang, S. Reliability Study of Zinc Oxide Thin-Film Transistor with High-K Gate Dielectric. In Proceedings of the 2012 IEEE International Conference on Electron Devices and Solid State Circuit (EDSSC), Bangkok, Thailand, 3–5 December 2012; pp. 1–2. [CrossRef]
6. Gao, H.-X.; Hu, R.; Yang, Y.-T. The Theoretical Investigation and Analysis of High-Performance ZnO Double-Gate Double-Layer Insulator Thin-Film Transistors. *Chin. Phys. Lett.* **2012**, *29*, 017305. [CrossRef]
7. Xie, J.; Zhu, Z.; Tao, H.; Zhou, S.; Liang, Z.; Li, Z.; Yao, R.; Wang, Y.; Ning, H.; Peng, J. Research Progress of High Dielectric Constant Zirconia-Based Materials for Gate Dielectric Application. *Coatings* **2020**, *10*, 698. [CrossRef]
8. Robertson, J. High Dielectric Constant Oxides. *Eur. Phys. J. Appl. Phys* **2004**, *28*, 265–291. [CrossRef]
9. Wang, Y.; Han, T.; Yang, L.; Wang, B.; Xue, J.; Yi, S.; Xie, A.; Cao, C. A Novel IGZO-TFT Device With High Dielectric Constant Sandwich Structure Gate Dielectric Layer. In Proceedings of the 2023 24th International Conference on Electronic Packaging Technology (ICEPT), Shihezi City, China, 8–11 August 2023; pp. 1–5. [CrossRef]
10. Shrivastava, S.; Bahubalindruni, P.G.; Goes, J. A Pulse Width Modulator Using a High-Speed Comparator With Flexible Oxide TFT Technology. *IEEE Solid-State Circuits Lett.* **2022**, *5*, 288–291. [CrossRef]
11. Deng, X.; Zhang, Y.; Fu, H.; Zhang, S. High Mobility Metal-Oxide Thin Film Transistors with IGZO/In_2O_3 Dual-Channel Structure. In Proceedings of the 2018 9th Inthernational Conference on Computer Aided Design for Thin-Film Transistors (CAD-TFT), Shenzhen, China, 16–18 November 2018; pp. 1–3. [CrossRef]
12. Yang, Z.; Pu, H.; Cui, C.; Zhang, L.; Dong, C.; Zhang, Q. Solution-Processed Indium–Zinc-Oxide Thin Film Transistors With High-k Magnesium Titanium Oxide Dielectric. *IEEE Electron. Device Lett.* **2014**, *35*, 557–559. [CrossRef]
13. Wang, B.; Huang, W.; Chi, L.; Al-Hashimi, M.; Marks, T.J.; Facchetti, A. High-k Gate Dielectrics for Emerging Flexible and Stretchable Electronics. *Chem. Rev.* **2018**, *118*, 5690–5754. [CrossRef]
14. Devaray, P.; Hatta, S.F.W.M.; Wong, Y.H. An Overview of Conventional and New Advancements in High Kappa Thin Film Deposition Techniques in Metal Oxide Semiconductor Devices. *J. Mater. Sci. Mater. Electron.* **2022**, *33*, 7313–7348. [CrossRef]
15. Johnson, R.W.; Hultqvist, A.; Bent, S.F. A Brief Review of Atomic Layer Deposition: From Fundamentals to Applications. *Mater. Today* **2014**, *17*, 236–246. [CrossRef]
16. Christen, H.M.; Eres, G. Recent Advances in Pulsed-Laser Deposition of Complex Oxides. *J. Phys. Condens. Matter* **2008**, *20*, 264005. [CrossRef]
17. Joung, S.-Y.; Yim, H.; Lee, D.; Shim, J.; Yoo, S.Y.; Kim, Y.H.; Kim, J.S.; Kim, H.; Hyeong, S.-K.; Kim, J.; et al. All-Solution-Processed High-Performance MoS_2 Thin-Film Transistors with a Quasi-2D Perovskite Oxide Dielectric. *ACS Nano* **2024**, *18*, 1958–1968. [CrossRef] [PubMed]
18. Jaehnike, F.; Pham, D.V.; Anselmann, R.; Bock, C.; Kunze, U. High-Quality Solution-Processed Silicon Oxide Gate Dielectric Applied on Indium Oxide Based Thin-Film Transistors. *ACS Appl. Mater. Interfaces* **2015**, *7*, 14011–14017. [CrossRef] [PubMed]
19. Ji, S.-B.; Seong, N.; Park, J.; Im, H.; Kim, Y.-S.; Hong, Y. Field-Effect Mobility Extraction of Solution-Processed InGaZnO Thin-Film Transistors Considering Dielectric Dispersion Behavior of AlO_x Gate Insulator. *ACS Appl. Electron. Mater.* **2023**, *5*, 1035–1040. [CrossRef]

20. He, W.; Wang, S.; Hang, B.; Wei, X.; Liang, L. Development of Solution-Processed Organic Semiconductor Thin Films. In *Interdisciplinary Research for Printing and Packaging*; Zhao, P., Ye, Z., Xu, M., Yang, L., Zhang, L., Yan, S., Eds.; Springer Singapore: Singapore, 2022; pp. 471–479. [CrossRef]
21. Liu, Q.; Zhao, C.; Zhao, T.; Liu, Y.; Mitrovic, I.Z.; Xu, W.; Yang, L.; Zhao, C.Z. Ecofriendly Solution-Combustion-Processed Thin-Film Transistors for Synaptic Emulation and Neuromorphic Computing. *ACS Appl. Mater. Interfaces* **2021**, *13*, 18961–18973. [CrossRef]
22. Quah, H.J.; Cheong, K.Y.; Hassan, Z.; Lockman, Z.; Jasni, F.A.; Lim, W.F. Effects of Postdeposition Annealing in Argon Ambient on Metallorganic Decomposed CeO_2 Gate Spin Coated on Silicon. *J. Electrochem. Soc.* **2009**, *157*, H6. [CrossRef]
23. Xu, W.; Long, M.; Zhang, T.; Liang, L.; Cao, H.; Zhu, D.; Xu, J.-B. Fully Solution-Processed Metal Oxide Thin-Film Transistors via a Low-Temperature Aqueous Route. *Ceram. Int.* **2017**, *43*, 6130–6137. [CrossRef]
24. Jo, J.-W.; Kim, K.-H.; Kim, J.; Ban, S.G.; Kim, Y.-H.; Park, S.K. High-Mobility and Hysteresis-Free Flexible Oxide Thin-Film Transistors and Circuits by Using Bilayer Sol–Gel Gate Dielectrics. *ACS Appl. Mater. Interfaces* **2018**, *10*, 2679–2687. [CrossRef]
25. Wang, S.; Yao, S.; Lin, J.; Xia, G. Eco-Friendly, Low-Temperature Solution Production of Oxide Thin Films for High-Performance Transistors via Infrared Irradiation of Chloride Precursors. *Ceram. Int.* **2019**, *45*, 9829–9834. [CrossRef]
26. Arias, A.C.; MacKenzie, J.D.; McCulloch, I.; Rivnay, J.; Salleo, A. Materials and Applications for Large Area Electronics: Solution-Based Approaches. *Chem. Rev.* **2010**, *110*, 3–24. [CrossRef] [PubMed]
27. Jung, Y.; Jun, T.; Kim, A.; Song, K.; Yeo, T.H.; Moon, J. Direct Photopatternable Organic–Inorganic Hybrid Gate Dielectric for Solution-Processed Flexible ZnO Thin Film Transistors. *J. Mater. Chem.* **2011**, *21*, 11879–11885. [CrossRef]
28. Hwang, J.; Lee, J.; Kim, Y.; Lee, E.; Wang, Y.; Kim, H. Hybrid Gate Insulator for OTFT Using Dip-Coating Method. *Curr. Appl. Phys.* **2011**, *11*, S154–S157. [CrossRef]
29. Wong, H.-S.P.; Lee, H.-Y.; Yu, S.; Chen, Y.-S.; Wu, Y.; Chen, P.-S.; Lee, B.; Chen, F.T.; Tsai, M.-J. Metal-Oxide RRAM. *Proc. IEEE* **2012**, *100*, 1951–1970. [CrossRef]
30. El-Shereafy, E.; Abousekkina, M.M.; Mashaly, A.; El-Ashry, M. Mechanism of Thermal Decomposition and γ-Pyrolysis of Aluminum Nitrate Nonahydrate [$Al(NO_3)_3 \cdot 9H_2O$]. *J. Radioanal. Nucl. Chem.* **1998**, *237*, 183–186. [CrossRef]
31. Melnikov, P.; Nascimento, V.A.; Arkhangelsky, I.V.; Zanoni Consolo, L.Z. Thermal Decomposition Mechanism of Aluminum Nitrate Octahydrate and Characterization of Intermediate Products by the Technique of Computerized Modeling. *J. Therm. Anal. Calorim.* **2013**, *111*, 543–548. [CrossRef]
32. Mandai, T.; Johansson, P. Haloaluminate-Free Cationic Aluminum Complexes: Structural Characterization and Physicochemical Properties. *J. Phys. Chem. C* **2016**, *120*, 21285–21292. [CrossRef]
33. Jung, S.H.; Han, H.S.; Kim, Y.B.; Kim, D.S.; Deshpande, N.G.; Oh, S.J.; Choi, J.H.; Cho, H.K. Toward Ultraviolet Solution Processed ZrO_x/IZO Transistors with Top-Gate and Dual-Gate Operation: Selection of Solvents, Precursors, Stabilizers, and Additive Elements. *J. Alloys Compd.* **2020**, *847*, 156431. [CrossRef]
34. Padhi, P.S.; Ajimsha, R.S.; Chetia, S.K.; Das, A.K.; Sahu, V.K.; Misra, P. Reduced Leakage Current in $Al_2O_3/TiO_2/Al_2O_3$ Dielectric Stacks Grown by Pulsed Laser Deposition. *AIP Conf. Proc.* **2020**, *2265*, 030190. [CrossRef]
35. Kwak, H.Y.; Kwon, H.M.; Kwon, S.K.; Jang, J.H.; Sung, S.Y.; Lim, S.; Lee, H.D. Dielectric Relaxation and Reliability of Al_2O_3-HfO_2-Al_2O_3 Sandwiched Metal-Insulator-Metal (MIM) Capacitor. *AMR* **2013**, *658*, 116–119. [CrossRef]
36. Choi, C.H.; Kim, T.; Ueda, S.; Shiah, Y.-S.; Hosono, H.; Kim, J.; Jeong, J.K. High-Performance Indium Gallium Tin Oxide Transistors with an Al_2O_3 Gate Insulator Deposited by Atomic Layer Deposition at a Low Temperature of 150 °C: Roles of Hydrogen and Excess Oxygen in the Al_2O_3 Dielectric Film. *ACS Appl. Mater. Interfaces* **2021**, *13*, 28451–28461. [CrossRef] [PubMed]

Disclaimer/Publisher's Note: The statements, opinions and data contained in all publications are solely those of the individual author(s) and contributor(s) and not of MDPI and/or the editor(s). MDPI and/or the editor(s) disclaim responsibility for any injury to people or property resulting from any ideas, methods, instructions or products referred to in the content.

micromachines

MDPI

Article

Capacitance–Voltage Fluctuation of Si_xN_y-Based Metal–Insulator–Metal Capacitor Due to Silane Surface Treatment

Tae-Min Choi, Eun-Su Jung, Jin-Uk Yoo, Hwa-Rim Lee and Sung-Gyu Pyo

School of Integrative Engineering, Chung-Ang University, 84, Heukseok-ro, Dongjak-gu, Seoul 06974, Republic of Korea; c79411@gmail.com (T.-M.C.); eunsuj@cau.ac.kr (E.-S.J.); wlsdnr5771@naver.com (J.-U.Y.); ghkfla0725@naver.com (H.-R.L.)
* Correspondence: sgpyo@cau.ac.kr

Abstract: In this study, we analyze metal–insulator–metal (MIM) capacitors with different thicknesses of SixNy film (650 Å, 500 Å, and 400 Å) and varying levels of film quality to improve their capacitance density. SixNy thicknesses of 650 Å, 500 Å, and 400 Å are used with four different conditions, designated as MIM (N content 1.49), NEWMIM (N content 28.1), DAMANIT (N content 1.43), and NIT (N content 0.30). We divide the C–V characteristics into two categories: voltage coefficient of capacitance (VCC) and temperature coefficient of capacitance (TCC). There was an overall increase in the VCC as the thickness of the SixNy film decreased, with some variation depending on the condition. However, the TCC did not vary significantly with thickness, only with condition. At the same thickness, the NIT condition yielded the highest capacitance density, while the MIM condition showed the lowest capacitance density. This difference was due to the actual thickness of the film and the variation in its k-value depending on the condition. The most influential factor for capacitance uniformity was the thickness uniformity of the SixNy film.

Keywords: MIM; capacitors; metal–insulator–metal; electrical performance; Si_xN_y; cap density; VCC; TCC

Citation: Tae-Min Choi, Eun-Su Jung, Jin-Uk Yoo, Hwa-Rim Lee and Sung-Gyu Pyo. Capacitance–Voltage Fluctuation of Si_xN_y-Based Metal–Insulator–Metal Capacitor Due to Silane Surface Treatment. *Micromachines* **2024**, *15*, 1204. https://doi.org/10.3390/mi15101204

Academic Editor: Chengyuan Dong

Received: 5 September 2024
Revised: 25 September 2024
Accepted: 27 September 2024
Published: 28 September 2024

Copyright: © 2024 by the authors. Licensee MDPI, Basel, Switzerland. This article is an open access article distributed under the terms and conditions of the Creative Commons Attribution (CC BY) license (https://creativecommons.org/licenses/by/4.0/).

1. Introduction

A metal–insulator–metal (MIM) capacitor is an analog integrated circuit (IC) configuration device with the advantages of low electrode resistance and parasitic capacitance [1–4]. MIM capacitors have high charge mobility and burst power characteristics that make them excellent energy-storage devices and potential auxiliary power sources.

IM capacitors have been applied to ICs such as high-power microprocessor units and dynamic random-access memory. However, with the development of wireless communication, their application to radio frequency (RF) devices has been actively studied [5–7]. As current RF devices require high operating frequencies, MIM devices also require high capacitance per unit area [8–11].

According to this demand, the design of the structure, thin-film deposition method, selection of the bottom and top electrode materials, insulator material, thickness of the electrode and insulator, dielectric constant of the insulator, and crystal structure of the insulator must be considered in depth to produce a high-performance MIM capacitor with a high capacitance [12–15]. In the evaluation of MIM capacitors, it is important to conduct a comprehensive analysis considering factors such as capacitance density (CD), leakage current density, charge storage density, and dielectric breakdown strength [16–19].

The choice of insulator material is a crucial factor in capacitors, and the CD relies heavily on the dielectric constant and thickness of the insulator. According to Equation (1), the CD increases when the dielectric constant is higher, which is a natural property of the insulator, and when the thickness is lower [1,20]:

$$C = \frac{k\varepsilon_0 A}{d} \rightarrow \frac{C}{A} = \frac{k\varepsilon_0}{d} \quad (1)$$

where C denotes the capacitance (F), k is the dielectric constant, ε_0 is the permittivity of the vacuum (8.854×10^{-12}/m), and d is the thickness of the insulator (m).

Because of the abovementioned reasons, it is generally necessary to introduce a high-k material to increase the dielectric constant [21,22], which entails considerable investment and time because it requires equipment, facilities, and source replacement. In addition, according to Natori et al., the relative permittivity of the material (k) decreases as the insulator thickness of the capacitor decreases, and k tends to decrease significantly for high-k materials [13,23]. Therefore, the introduction of high-k materials is subject to many limitations. However, if the thickness of the currently used medium-k dielectric material, Si_xN_y (k = 7), can be reduced by considering the leakage aspect, the capacitance value can be increased without using a high-k material [24–26]. Yu et al. explained the performance of HfO_2-based MIM capacitors deposited by the atomic layer deposition (ALD) method with respect to the thickness of the dielectric [27]. As the thickness of the HfO_2 insulator layer decreased, the CD and voltage coefficient of capacitance (VCC) increased [28–31]. In practice, GaAs-based MIM capacitors have been used in the past. In fact, for GaAs-based MIM capacitors, Si_xN_y is the most commonly applied material owing to its excellent electrical properties, compliant dielectric constant, high dielectric breakdown voltage, and low leakage current [28,32,33]. Moreover, the electrical properties can be improved by optimizing the deposition condition of Si_xN_y, which has the greatest effect on the electrical properties of MIM capacitors. Yota et al. confirmed that the stress, CD, breakdown voltage, and performance of MIM capacitors exhibited significant differences in each insulator layer, with a single layer or multiple layers of silicon nitride formed depending on the deposition conditions [34]. Therefore, the electrical properties of MIM capacitors can be improved by optimizing the deposition condition of Si_xN_y.

In this study, to develop an optimal condition for the deposition conditions of Si_xN_y that improves the insulator properties of MIM capacitors and secures feasibility, we fabricated MIM capacitors with different thicknesses of Si_xN_y and deposition conditions on M4 wiring and then evaluated the capacitance–voltage (C–V) characteristics, focusing on the evaluation of cap density uniformity, the dielectric temperature coefficient of capacitance (TCC), and the VCC.

2. Materials and Methods

Patterned 200 mm Si (100) wafers were used to measure the integration process steps. Several different cap dielectrics were investigated and deposited by PE-ALD. An Applied Materials MIRRA tool(Applied Materials, Gloucester, MA, USA) was used for the blanket. In this study, the density of MIM capacitors was considered to be 8 fF/m^2, and the capacitors were fully integrated using the 0.15 m Al interconnect processes. The first single MIM capacitor was formed using metal 3 and metal 4 to minimize the effect of the parasitic coupling of the silicon substrates. In addition, the second single MIM capacitor was formed using metal 5 and metal 6. It is crucial for MIM capacitors to have a symmetric structure by having identical boundary conditions on both sides of the dielectric [8,35,36].

The bottom electrode of the MIM capacitor was prepared using Ti (100 Å)/Al–Cu (4500 Å)/Ti (50 Å)/TiN (600 Å) wiring, and the top electrode was prepared using TiN (1500 Å). The insulator was SixNy. The capacitor fabrication process was as follows: bottom electrode deposition → bottom electrode scrub → insulator deposition → top metal deposition → MIM PH → MIM → TOP METAL etching → (((CH_3)4NOH:H_2O) cleaning 1 → MIM asher → (((CH_3)4NOH:H_2O) cleaning 2 → insulator etching → ACT 935 (wet PR strip solution including amine) →UVAS. MIM ET was performed using the endpoint detection method.

The top electrode layer was connected to the upper metal layer through a dense matrix of vias. All wafers mentioned in this paper were passivated using PE-ALD nitride and were annealed below 450 °C. To improve the voltage linearity, Interface plasma treatment was administered before and after the PE-ALD dielectrics. Further details of the interface plasma treatment and thickness ratio in each stack layer are listed in Table 1. The blanket film characteristics of PE-ALD dielectrics were evaluated by using an ellipsometer at 673 nm to measure the thickness, refractive index, and uniformity of the PE-ALD dielectrics. A Hg probe was used to measure the dielectric constant, and the deposition rate was calculated according to the thickness slope as a function of the cycle times [4].

Table 1. (**a**) Si_xN_y film properties and (**b**) corresponding process conditions.

(a)	MIM	NEW MIM 650	DAMA NIT	NIT
Dep. rate	~149 Å/s	~29 Å/s	~59 Å/s	88 Å/s
Within W/F unit (1σ)	1.14%	1.90%	2.34%	2.77%
W/F to W/F unit (1σ)	1.58%	2.21%	1.05%	2.30%
Stress	-2.23×10^9	-1.75×10^{10}	-2.34×10^9	
H content (N-H: Si-H)	12.7%: 8.5%	22.5%: 0.8%	10.5%: 7.3%	4.4%: 14.8%
N content (N-H/Si-H)	1.49	28.1	1.43	0.30
(b)	**MIM 650**	**NEW MIM 650**	**DAMA NIT**	**NIT 650**
Step end control	By time	By time	By time	By time
Maximum step time	4.4 s	22.8 s	11.0 s	
Endpoint selection	No endpoint	No endpoint	No endpoint	No endpoint
Pressure	Servo 4.25 Torr	Servo 4.25 Torr	Servo 4.2 Torr	Servo 4.5 Torr
RF power	690 W	690 W	420 W	425 W
Susc. temperature	400 °C	400 °C	400 °C	400 °C
Susceptor spacing	620 mils	620 mils	550 mils	475 mils
N_2	3800 sccm	3800 sccm	2500 sccm	4000 sccm
NH_3	130 sccm	50 sccm	38 sccm	60 sccm
SiH_4	260 sccm	100 sccm	110 sccm	170 sccm

The C–V characteristics were measured manually using an LCR meter (HP4284A, Agilent, Santa Clara, CA, USA) under the conditions given in Table 2. The VRDB was performed based on the JESD35-A standard. The thickness analysis of Si_xN_y per condition was performed using transmission electron microscopy (CM200FEGTEM)/scanning transmission electron microscopy (STEMoperated at 300 keV with an energy-dispersive X-ray spectroscopy (EDS) SUTW-SiLi X-ray detector and a Gatan 666 parallel electron energy loss spectroscopy (PEELS) spectrometer (Philips, Eindhoven, Netherlands), and a focused ion beam (FIB). The via resistance and via chain yields were measured in dual-damascene structures. A wafer-level bias thermal stress (BTS) test was performed under different conditions to verify the effectiveness of the barrier layers. Failures were analyzed by scanning electron microscopy (SEM) and X-ray spectroscopy (EDX).(SIGMA, Carl Zeiss, Jena, Germany)

Table 2. C–V characterization measurement conditions.

Parameter	Setting
Display mode	Cp (parallel capacitor), D (dissipation factor)
Sweep voltage (V)	−5~5
Step (V)	0.5
Oscillation	0.025
Frequency (kHz)	100
Capacitor size (μm²)	10 × 10, 15 × 15, 20 × 20, 25 × 25, 30 × 30, 50 × 50
Measurement points	Three points (top, center, and bottom)
Temperature (°C)	25, 50, 75, 100, 125

3. Results and Discussion

To understand the wafer-wide trend in CD, PCM measurements were performed with the split conditions given in Table 1; the corresponding results for a 25 × 25 cap size are presented in Figure 1.

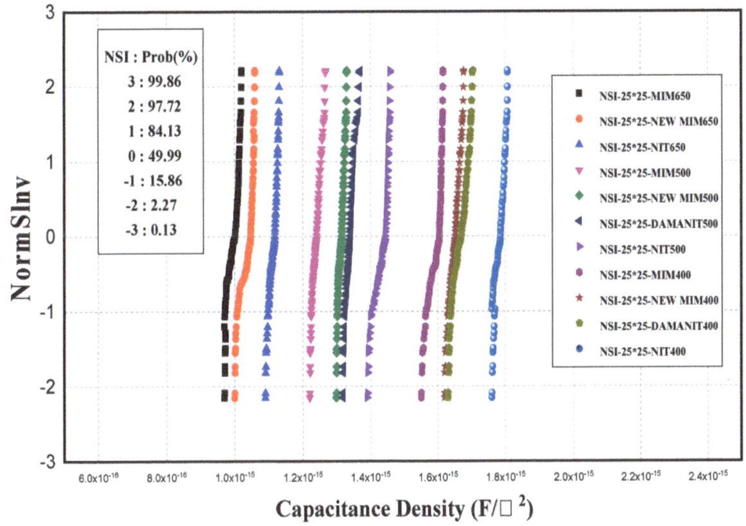

Figure 1. Accumulation curves for CD obtained from different thicknesses and conditions of SixNy films for MIM capacitors.

The CD varied depending on the split condition but was uniform within the wafer. For all conditions, the CD increased with decreasing thickness, and at the same thickness, it varied slightly between conditions. By examining the range of CD in relation to thickness, the following values were observed for thicknesses of 650, 500, and 400 Å, respectively: 0.983–1.1, 1.24–1.4, and 1.57–1.79 fF/μm^2. The difference in capacitance densities between conditions at the same thickness can be considered to be the difference between the actual thickness of the Si_xN_y film and the target thickness and the difference in the k-value of the deposited film by condition. To confirm this, 500 Å thick Si_xN_y films deposited under the different conditions were analyzed by TEM; the corresponding results are shown in Figure 2.

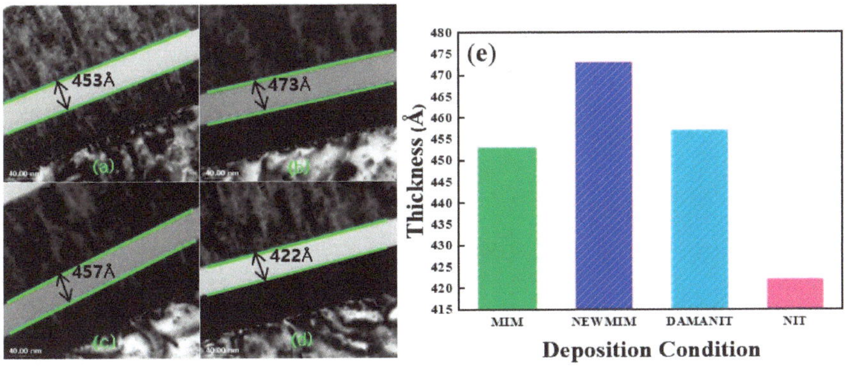

Figure 2. Cross-sectional TEM images of Si_xN_y films with (**a**) MIM, (**b**) NEWMIM, (**c**) DAMANIT, and (**d**) NIT. The thicknesses of the Si_xN_y films are presented in the (**e**) rod graph.

Figure 2a–d depict the cross-sectional TEM images of Si_xN_y films deposited under MIM, NEWMIM, DAMANIT, and NIT conditions, respectively, and Figure 2e shows a barplot of the thickness of Si_xN_y films obtained from the TEM images. The NIT condition showed the lowest thickness, while the NEWMIM condition showed the highest thickness. For accurate analysis, the k-value was calculated after matching the TEM analysis die and the PCM measurement die; the corresponding results are presented in Table 3.

Table 3. The k-values of the MIM capacitor with Si_xN_y conditions of MIM, NEWMIM, DAMANIT, and NIT.

Deposition Condition	MIM	NEWMIM	DAMANIT	NIT
CD (fF/μm^2)	1.3256	1.3203	1.3103	1.3606
Thickness (TEM, Å)	453	473	457	422
k-value ($\varepsilon_0 \cdot \varepsilon$)	6.00×10^{-17}	6.25×10^{-17}	5.99×10^{-17}	5.74×10^{-17}

In this experiment, we compared the capacitance densities of four materials with the same thickness (500 Å) and found that NIT afforded the highest CD, followed by NEWMIM, DAMANIT, and MIM. However, we noticed that the deposited thickness did not follow this trend. This indicates that thickness alone is not the only factor that affects the CD. The difference in the k-value according to the condition also appears to play a significant role in determining the CD. In general, the k-value is influenced by two primary factors, namely, the macroscopic electric field and the dipole moment per unit volume, as given by Equation (2) [2]:

$$K = 1 + \frac{4\pi P}{E} \tag{2}$$

where P is the dipole moment per unit volume and E is the macroscopic electric field.

Because the p-value is dependent on the electronic polarizability, it is affected by the bond conformation and bond strength [21]. The Si_xN_y films had different dipole moments due to the different values of Si-H/N-H (Table 4) depending on the deposition condition; therefore, the k-value was different for each condition.

Table 4. Properties of SixNy films according to different conditions.

	MIM 650 DEP	NEW MIM 650 DEP	DAMA NIT 650 DEP	NIT 650 DEP
Deposition rate (Å/s)	~149	~29	~59	88
Within W/F unit (1σ, %)	1.14	1.90	2.34	2.77
W/F to W/F unit (1σ, %)	1.58	2.21	1.05	2.30
Stress	-2.23×10^9	-1.75×10^{10}	-2.34×10^9	-
H content (N-H:Si-H)	12.7%:8.5%	22.5%:0.8%	10.5%:7.3%	4.4%:14.8%

To check the variation in CD according to capacitor size, the capacitor density by thickness and condition was measured for 10 × 10, 15 × 15, 20 × 20, 25 × 25, 30 × 30, and 50 × 50 μm^2 samples; it is plotted in Figure 3. In this case, the CD according to size was taken as the average value within the wafer.

It can be observed that the CD decreases as the size of the capacitor increases, regardless of the thickness and condition of Si_xN_y. After a certain point, the CD remains constant. This phenomenon can be attributed to the effect of fringe capacitance due to the perimeter/area ratio and the variation in fringe impedance CD (FICD) with respect to size. The difference between capacitor sizes of 10 × 10 and 20 × 20 μm^2 is more pronounced in the case of FICD variation, as it has a greater impact on smaller sizes [37].

In terms of device fabrication, the uniformity of the CD is closely related to the process capability index (Cp, Cpk), with Si_xN_y thickness uniformity being the most important factor.

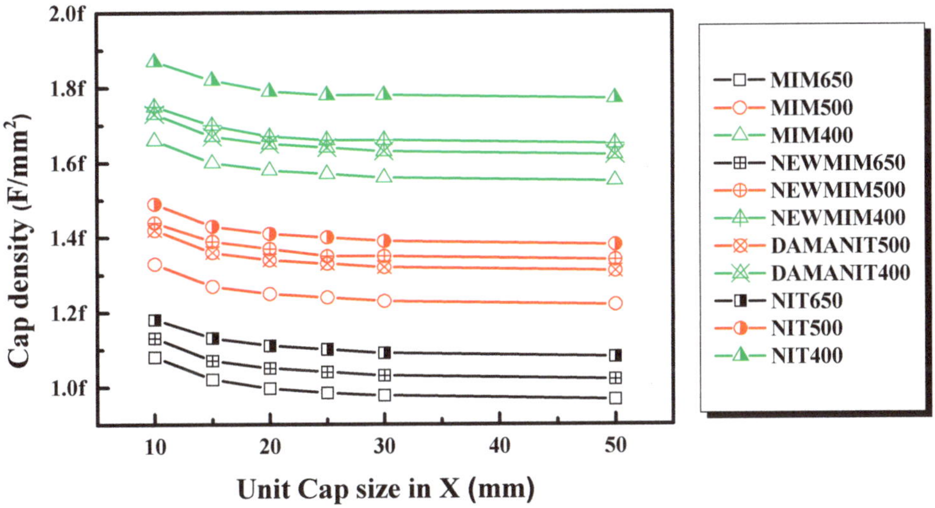

Figure 3. Effect of capacitor size on MIM CD with different Si_xN_y thicknesses and conditions.

As depicted in Figure 4, although there are a few points that deviate from the linear trend, a proportional relationship exists between Si_xN_y thickness nonuniformity and CD nonuniformity, with a slope of 1.04.

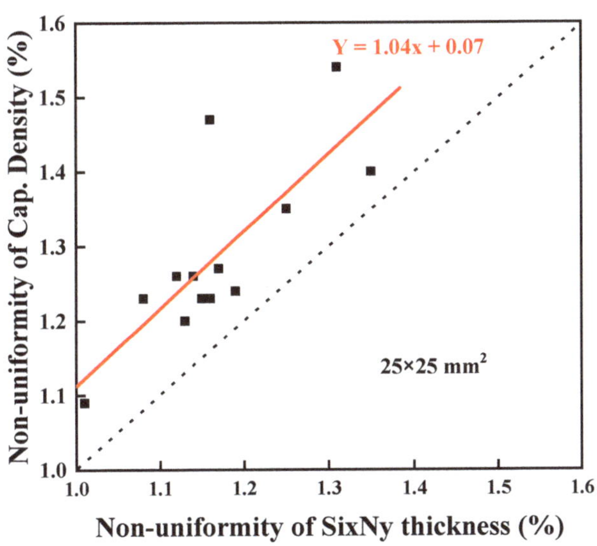

Figure 4. Relationship between nonuniformity of CD and Si_xN_y thickness.

Improving the uniformity of Si_xN_y thickness can result in an improvement in CD uniformity, which, in turn, can increase the values of Cp and Cpk on the device manufacturing side to 1.33 or above.

To measure the VCC, an index indicating the degree of change in capacitance with respect to voltage variations, measurements were taken at the top, center, and bottom of the wafers according to the thickness of Si_xN_y and the conditions. The VCC is denoted by V_{cc1} (ppm/dV) and V_{cc2} (ppm/dV^2), as expressed by Equation (3) [38]:

$$\frac{C(V) - C(0)}{C(0)} = V_{cc2}V^2 + V_{cc1}V + C \tag{3}$$

where $C(V)$ is the capacitance under variable voltage, $C(0)$ is the capacitance at 0 V, V_{cc1} and V_{cc2} are the VCCs, and C is a constant value.

The VCC graph was plotted by performing a polynomial fit with the voltage on the X-axis and the normalized ΔC on the y-axis, as described in Equation (3). As an example, the VCC graph for the Si_xN_y film processed with the NEWMIM condition at a thickness of 500 Å is depicted in Figure 5.

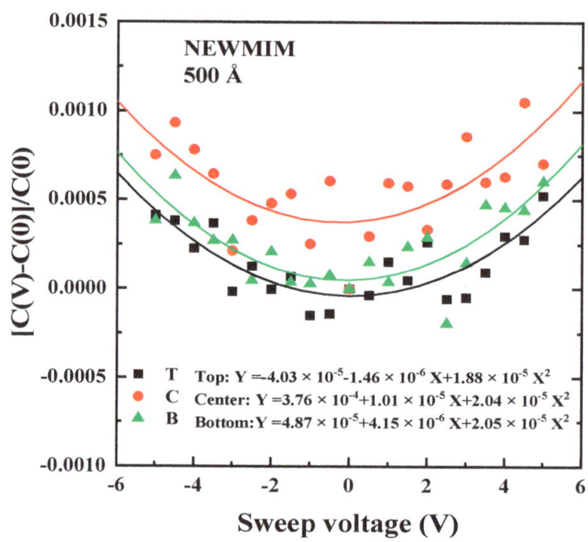

Figure 5. Normalized ΔC vs. voltage at the top, center, and bottom positions. The condition and thickness of the Si_xN_y film are NEWMIM and 500 Å, respectively.

The VCC graph results for the Si_xN_y films, which vary in thickness (650 Å, 500 Å, and 400 Å) and condition (MIM, NEWMIM, DAMANIT, and NIT), are summarized in Figure 6. Figure 6a displays the V_{cc1} values according to thickness, while Figure 6b illustrates the V_{cc2} values as a function of thickness. Both the V_{cc1} and V_{cc2} values showed an increasing trend as the thickness decreased, with the initial level and degree of increase varying according to the condition.

In the case of V_{cc1}, all conditions showed values below 60 ppm/dV at 650 Å, but they values increased as the thickness decreased, and only the MIM and NEWMIM conditions showed values over 60 ppm/V. V_{cc2} tended to increase as the thickness decreased; however, all other conditions except NIT could satisfy the value of 100 ppm/dV^2 or less when implementing a 2-fF/µm^2 MIM capacitor.

To investigate the TCC characteristics of the MIM capacitors, the capacitance was measured at the center of the wafers with varying thicknesses and conditions of the Si_xN_y films while incrementally raising the temperature to 25 °C, 50 °C, 75 °C, 100 °C, and 125 °C. The TCC was calculated using Equation (4) [39]:

$$\frac{C(T) - C(25)}{C(25)} = T_{cc}T + C \tag{4}$$

where $C(T)$ is the capacitance under variable temperature, $C(25)$ is the capacitance at 25 °C, T_{CC} is the TCC, and C is a constant value.

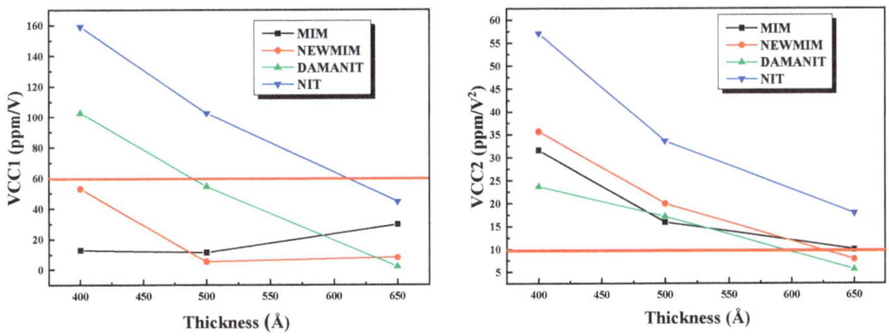

(a) (b)

Figure 6. Graphs of (**a**) VCC1 and (**b**) VCC2 vs. thickness with Si_xN_y conditions.

The TCC serves as an indicator of the degree of change in capacitance in response to temperature variations. Unlike the VCC, the TCC exhibits a linear relationship with temperature. Therefore, when the temperature is plotted on the X-axis and the normalized ΔC on the Y-axis, the slope value corresponds to the TCC value.

As an illustration, a TCC graph for Si_xN_y films fabricated with the NEWMIM condition at thicknesses of 650, 500, and 400 Å is presented in Figure 7. Additionally, the results for the other conditions are presented to depict the variation in the TCC values according to thickness in Figure 8. Except for the NIT condition, the remaining conditions exhibited values below 50 ppm/dT, even as the thickness decreased. Moreover, the TCC values varied according to the condition at identical thicknesses.

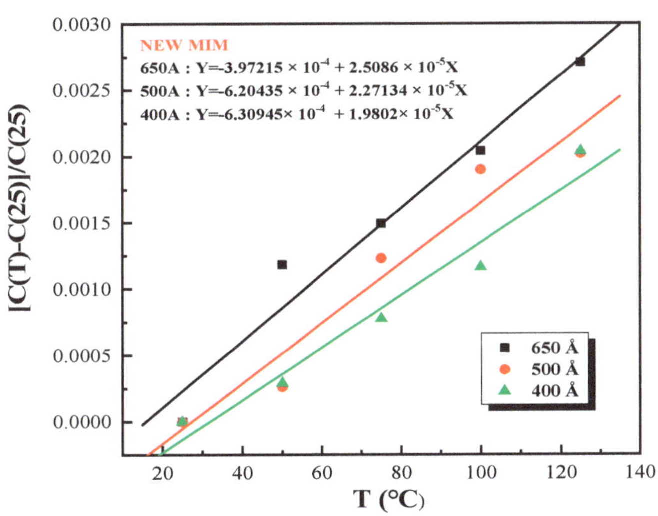

Figure 7. The TCC graph for SixNy films fabricated with the NEWMIM condition at thicknesses of 650, 500, and 400 Å.

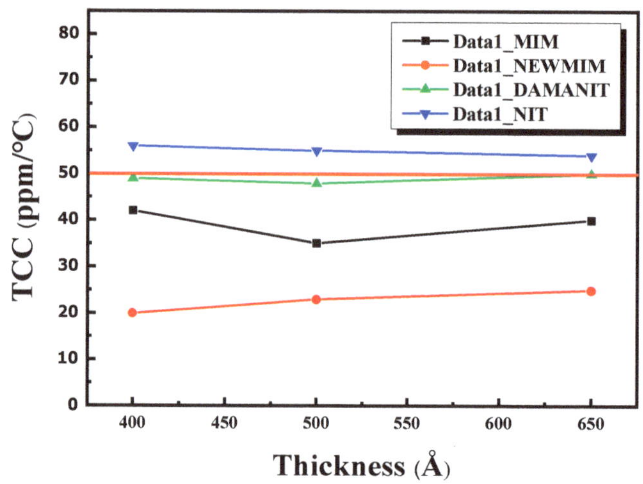

Figure 8. Graph of TCC1 vs. thickness with different Si_xN_y conditions.

As revealed by Table 4, the differences in Si_xN_y conditions are attributed to the N-H/Si-H ratio. To depict the changes in the TCC due to film quality, the N-H/Si-H vs. TCC values are plotted in Figure 9. The results indicate that as the N-H/Si-H ratio increased, the TCC values exhibited an exponential decay trend, confirming that the TCC values are influenced by the quality of the Si_xN_y film.

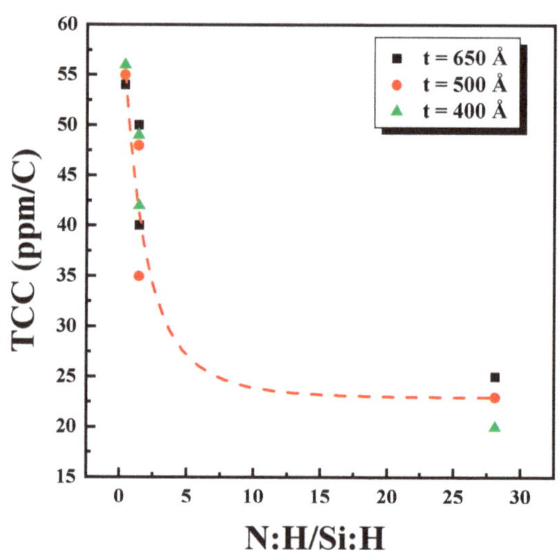

Figure 9. Graph of TCC vs. thickness with varying N:H/Si-H ratio.

4. Conclusions

An evaluation of the C–V characteristics was conducted for MIM capacitors based on the insulator (Si_xN_y) deposition thickness and deposition conditions. The CD values were in the ranges of 0.983–1.1, 1.24–1.4, and 1.57–1.79 fF/μm^2 for 650, 500, and 400 Å, respectively. Further, the CD increased as the thickness decreased, with variations across different conditions.

At the same thickness, the NIT condition exhibited the highest CD, while the MIM condition showed the lowest. This discrepancy is attributed to the effect of the actual thickness angle for incident light. Through this effect, incident angles can be obtained according to the changing rules of the self-images [10]. However, angle sensors that rely on lenses or gratings are limited by diffraction theory [11]. When a sensor's scale is close to or smaller than the wavelength of the detection light, its accuracy will sharply decrease due to the minimal recognizable half wavelength limits between two adjacent imaged spots, and this makes it difficult to achieve the design goal of ultra-thin sensors with deep miniaturization.

As for resonance coupling, it typically occurs in sub-wavelength structures or devices such as nanowires and surface plasmon, it describes the interaction of structures under excitation, and its characteristics are tightly related to the incentives. Through elaborate design and testing, the resonance can be used for angle detection, which can be deduced by measurable physical quantities, e.g., photocurrents [12,13]. And to address the challenge of miniaturization on light-direction sensors, inspired by the directional sensing ability of small animals to sound and light waves, angle sensors based on coupled nanowires have demonstrated the high-precision angle perception of incident light at sub-wavelength scales and can be manufactured with existing mature nanofabrication processes [14–17].

Perovskite nanocrystals have excellent optoelectronic properties and can produce adjustable emissions with high color saturation under specific spectra, which can convert light-field information into color output. By arranging multicolor-emitting perovskite nanocrystals geometrically, the incident light angles can be calculated by decoding the color output, which utilizes the mutual occlusion between the nanocrystals under angular irradiation [18].

However, the above-mentioned incident light angle measurement methods are limited to planar angle measurement and belong to the two-dimensional (2D) level. Comprehensive detection of three-dimensional (3D) light-direction, including the elevation angle (θ) and azimuth angle (ϕ), typically relies on a combination of sensors, which often entails a compromise between a sensor's detection ability and physical size. Although the perovskite nanocrystal-based method is capable of sensing 3D light direction, its large-scale application is relatively costly and complex, and exploring new design mechanisms is necessary.

In order to overcome the miniaturization dilemma caused by diffraction limits and achieve comprehensive detection of 3D light-direction angles, in this work, we propose a 3D light-direction sensor unit based on the angular dependency of the resonant energy distribution within segmented concentric nanorings, capable of detecting incident light-direction upon a single device. Furthermore, a deep learning (DL) method is employed to fit and reason the calculation relationship between the angles and the proposed structural responses; thus, the trained data-driven model can be used to obtain angles. Additionally, the fabrication and measurement feasibility of the proposed sensor are shown. In terms of application, the proposed detector can be integrated with devices such as lenses and waveguides to enhance application flexibility and improve detection performance.

2. Materials and Methods

2.1. Nanoring Resonance for Angle Sensing

Paired resonant nanowires with a high refractive index have been theoretically and experimentally demonstrated to exhibit the strong angular sensitiveness of leaky-mode resonance, deeming are suitable for angle sensors [12]. Similarly, as a commonly used coupled waveguide device, concentric nanorings can couple and resonate with 3D light direction angular dependence under incident light irradiation [19]. For the concentri silicon nanorings shown in Figure 1a, nanorings with the same width (w) and height (h) ar arranged concentrically with a spacing (d), and the radius of the inner nanoring is define as r.

When space light is incident with the 3D direction of (θ, ϕ), optical Mie resonances a supported within the inner and outer nanoring cross-sections $a_i(\alpha)$ and $a_o(\alpha)$ determine

and the difference in the k-value of the Si_xN_y film according to the condition. Additionally, the CD was observed to decrease with increasing capacitor size, possibly due to the influence of fringe capacitance, which increased in proportion to the perimeter/area ratio.

The thickness uniformity of Si_xN_y was found to be the most significant factor affecting capacitance uniformity. Improvements in thickness uniformity can enhance Cp and Cpk on the device side. Across all conditions, a general increase was observed in the VCC as the thickness decreased, although there were some variations between conditions. However, the TCC showed no significant difference with thickness, indicating that the variations were mainly due to the conditions.

In summary, from the perspective of C–V analysis, all conditions, except NIT, demonstrated superior characteristics. Implementing thin Si_xN_y film depositions with stable uniformity using conditions other than NIT could potentially provide MIM capacitors with CD values of less than 100 ppm/dV^2, aiming for the achievement of 2 fF/μm^2.

Author Contributions: Conceptualization, T.-M.C., E.-S.J. and S.-G.P.; methodology, S.-G.P.; validation, J.-U.Y. and H.-R.L.; writing—original draft preparation, T.-M.C.; supervision, S.-G.P.; project administration, S.-G.P. All authors have read and agreed to the published version of the manuscript.

Funding: This work was supported by the Technology Innovation Program (or Industrial Strategic Technology Development Program—Public–private joint investment semiconductor R&D program (K-CHIPS) to foster high-quality human resources) ("RS-2023-00237003", High selectivity etching technology using cryoetch) funded by the Ministry of Trade, Industry and Energy (MOTIE, Republic of Korea) (1415187674).

Data Availability Statement: The original contributions presented in this study are included in the article. Further inquiries can be directed to the corresponding author.

Acknowledgments: This research was supported by the Chung-Ang University Graduate Research Scholarship in 2023.

Conflicts of Interest: The authors declare no conflicts of interest.

References

1. Xiong, L.; Hu, J.; Yang, Z.; Li, X.; Zhang, H.; Zhang, G. Dielectric Properties Investigation of Metal–Insulator–Metal (MIM) Capacitors. *Molecules* **2022**, *27*, 3951. [CrossRef] [PubMed]
2. Karthik, R.; Manjusha, K.A. Metal Insulator Metal Capacitors-State of the Art. *J. Comput. Theor. Nanosci.* **2018**, *15*, 2346–2349. [CrossRef]
3. Ding, S.-J.; Hu, H.; Zhu, C.; Kim, S.J.; Yu, X.; Li, M.-F.; Cho, B.J.; Chan, D.S.H.; Yu, M.B.; Rustagi, S.C.; et al. RF, DC, and reliability characteristics of ALD HfO_2-Al_2O_3 laminate MIM capacitors for Si RF IC applications. *IEEE Trans. Electron Devices* **2004**, *51*, 886–894. [CrossRef]
4. Choi, E.; Kim, A.; Kwon, S.H.; Pyo, S.G. Effect of Interface Treatment on the Voltage Linearity in 8 fF/μm^2 High-k Dielectric and Combination Stacks on Metal Insulator Metal (MIM) Capacitor. *Sci. Adv. Mater.* **2018**, *10*, 467–470. [CrossRef]
5. Hoa, P.T. High-K Dielectrics in Metal Insulator Metal (MIM) Capacitors for RF Applications. Ph.D. Thesis, National University of Singapore, Singapore, 2011.
6. Pavunny, S.P.; Misra, P.; Scott, J.F.; Katiyar, R.S. Advanced high-k dielectric amorphous $LaGdO_3$ based high density metal-insulator-metal capacitors with sub-nanometer capacitance equivalent thickness. *Appl. Phys. Lett.* **2013**, *102*, 252905. [CrossRef]
7. Ding, S.-J.; Hu, H.; Lim, H.; Kim, S.; Yu, X.; Zhu, C.; Li, M.; Cho, B.J.; Chan, D.S.; Rustagi, S.C. High-performance MIM capacitor using ALD high-k HfO_2-Al_2O_3 laminate dielectrics. *IEEE Electron Device Lett.* **2003**, *24*, 730–732. [CrossRef]
8. Sul, W.S.; Pyo, S.G. RF Characteristic Analysis Model Extraction on the Stacked Metal–Insulator–Metal Capacitors for Radio Frequency Applications. *IEEE Trans. Electron Devices* **2014**, *61*, 3011–3013. [CrossRef]
9. Mu, J.; Chou, X.; Ma, Z.; He, J.; Xiong, J. High-Performance MIM Capacitors for a Secondary Power Supply Application. *Micromachines* **2018**, *9*, 69. [CrossRef]
10. Li, W.; Lu, X.; Yang, R.; Liang, F.; Chen, W.; Xie, Z.; Zheng, J.; Zhu, J.; Huang, Y.; Yue, W.; et al. Highly sensitive and reproducible SERS substrates with binary colloidal crystals (bCCs) based on MIM structures. *Appl. Surf. Sci.* **2022**, *597*, 153654. [CrossRef]
11. Jeon, S.; Sung, S.-K.; Jang, E.-H.; Jeong, J.; Surabhi, S.; Choi, J.-H.; Jeong, J.-R. Multilayer metal-oxide-metal nanopatterns via nanoimprint and strip-off for multispectral resonance. *Appl. Surf. Sci.* **2018**, *428*, 280–288. [CrossRef]
12. Smitha, P.S.; Babu, V.S.; Shiny, G. Critical parameters of high performance metal-insulator-metal nanocapacitors: A review. *Mater. Res. Express* **2019**, *6*, 122003. [CrossRef]
13. Sejas-García, S.C.; Torres-Torres, R.; Valderrama-B, R.; Molina, J. Complex Permittivity Determination of Thin-Films through RF-Measurements of a MIM Capacitor. *IEEE Microw. Wirel. Compon. Lett.* **2014**, *24*, 805–807. [CrossRef]

14. Nam, M.; Kim, A.; Kang, K.; Choi, E.; Kwon, S.H.; Lee, S.J.; Pyo, S.G. Characterization of atomic layer deposited Al_2O_3/HfO_2 and Ta_2O_5/Al_2O_3 combination stacks. *Sci. Adv. Mater.* **2016**, *8*, 1958–1962. [CrossRef]
15. Jin Chung, J.; Hyuk Kim, T.; Ahsan Saeed, M.; Won Shim, J. Laminated indium-oxide/molybdenum-oxide nanocomposites for high-work-function electrodes in organic photovoltaics and capacitor devices. *Appl. Surf. Sci.* **2023**, *610*, 155526. [CrossRef]
16. Guo, Y.; Wang, S.; Du, X.; Liang, S.; Huang, S.; Peng, S.; Xie, Y.; Ma, M.; Xiong, L. Construction of ultrahigh capacity density carbon nanotube based MIM capacitor. *Energy Storage Mater.* **2023**, *63*, 103064. [CrossRef]
17. Ruhl, G.; Lehnert, W.; Lukosius, M.; Wenger, C.; Baristiran Kaynak, C.; Blomberg, T.; Haukka, S.; Baumann, P.K.; Besling, W.; Roest, A.; et al. Dielectric Material Options for Integrated Capacitors. *ECS J. Solid State Sci. Technol.* **2014**, *3*, N120–N125. [CrossRef]
18. Zheng, G.; He, Y.L.; Zhu, B.; Wu, X.; Zhang, D.W.; Ding, S.J. Improvement of Voltage Linearity and Leakage Current of MIM Capacitors With Atomic Layer Deposited Ti-Doped ZrO_2 Insulators. *IEEE Trans. Electron Devices* **2023**, *70*, 3064–3070. [CrossRef]
19. Park, Y.S.; Cho, S.-J.; Boo, J.-H.; Hong, B. Surface and electrical properties of organic–inorganic hybrid structure as gate insulator to organic thin film transistor. *Appl. Surf. Sci.* **2009**, *256*, 1023–1027. [CrossRef]
20. Wu, Y.-H.; Kao, C.-K.; Chen, B.-Y.; Lin, Y.-S.; Li, M.-Y.; Wu, H.-C. High density metal-insulator-metal capacitor based on $ZrO_2/Al_2O_3/ZrO_2$ laminate dielectric. *Appl. Phys. Lett.* **2008**, *93*, 033511. [CrossRef]
21. Xu, T.; Tekes, C.; Degertekin, F.L. CMUTs with high-K atomic layer deposition dielectric material insulation layer. *IEEE Trans. Ultrason. Ferroelectr. Freq. Control.* **2014**, *61*, 2121–2131. [CrossRef]
22. Sudheendran, K.; Pamu, D.; Ghanashyam Krishna, M.; James Raju, K.C. Determination of dielectric constant and loss of high-K thin films in the microwave frequencies. *Measurement* **2010**, *43*, 556–562. [CrossRef]
23. Holden, K.E.K.; Hall, G.D.R.; Cook, M.; Kendrick, C.; Pabst, K.; Greenwood, B.; Daugherty, R.; Gambino, J.P.; Allman, D.D.J. Dielectric Relaxation, Aging and Recovery in High-K MIM Capacitors. In Proceedings of the 2021 IEEE International Reliability Physics Symposium (IRPS), Monterey, CA, USA, 21–25 March 2021; pp. 1–10.
24. Zhao, C.; Zhao, C.Z.; Werner, M.; Taylor, S.; Chalker, P. Dielectric relaxation of high-k oxides. *Nanoscale Res. Lett.* **2013**, *8*, 456. [CrossRef]
25. Ohshima, I.; Cheng, W.; Ono, Y.; Higuchi, M.; Hirayama, M.; Teramoto, A.; Sugawa, S.; Ohmi, T. Reliability of silicon nitride gate dielectrics grown at 400 °C formed by microwave-excited high-density plasma. *Appl. Surf. Sci.* **2003**, *216*, 246–251. [CrossRef]
26. Ma, T.P. Gate dielectric properties of silicon nitride films formed by jet vapor deposition. *Appl. Surf. Sci.* **1997**, *117–118*, 259–267. [CrossRef]
27. Yu, H.Y.; Li, M.F.; Kwong, D.L. ALD $(HfO_2)_x(Al_2O_3)_{1-x}$ high-k gate dielectrics for advanced MOS devices application. *Thin Solid Films* **2004**, *462–463*, 110–113. [CrossRef]
28. Lue, H.T.; Lai, S.C.; Hsu, T.H.; Du, P.Y.; Wang, S.Y.; Hsieh, K.Y.; Liu, R.; Lu, C.Y. Understanding barrier engineered charge-trapping NAND flash devices with and without high-K dielectric. In Proceedings of the 2009 IEEE International Reliability Physics Symposium, Montreal, QC, Canada, 26–30 April 2009; pp. 874–882.
29. Cockbain, A.G.; Harrop, P.J. The temperature coefficient of capacitance. *J. Phys. D Appl. Phys.* **1968**, *1*, 1109. [CrossRef]
30. Khaldi, O.; Jomni, F.; Gonon, P.; Mannequin, C.; Yangui, B. Investigation of electrical properties of HfO_2 metal–insulator–metal (MIM) devices. *Appl. Phys. A* **2014**, *116*, 1647–1653. [CrossRef]
31. Lee, A.J.; Kim, B.S.; Hwang, J.H.; Kim, Y.; Oh, H.; Park, Y.; Jeon, W. Controlling the crystallinity of HfO_2 thin film using the surface energy-driven phase stabilization and template effect. *Appl. Surf. Sci.* **2022**, *590*, 153082. [CrossRef]
32. Li, H.; Yun, H.; Liang, W.; Dong, A.; Miao, M.; Sundaram, K.B. Characterization of Dielectric Breakdown and Lifetime Analysis for Silicon Nitride Metal-Insulator-Metal Capacitors under Electrostatic Discharge Stresses. In Proceedings of the 2018 IEEE International Symposium on the Physical and Failure Analysis of Integrated Circuits (IPFA), Singapore, 16–19 July 2018; pp. 1–5.
33. Kim, K.H.; Kim, K.S.; Ji, Y.J.; Kang, J.E.; Yeom, G.Y. Silicon nitride deposited by laser assisted plasma enhanced chemical vapor deposition for next generation organic electronic devices. *Appl. Surf. Sci.* **2021**, *541*, 148313. [CrossRef]
34. Yota, J. Effects of Deposition Method of PECVD Silicon Nitride as MIM Capacitor Dielectric for GaAs HBT Technology. *ECS Trans.* **2011**, *35*, 229. [CrossRef]
35. Iversen, C.-R. A high density MIM capacitor in a standard CMOS process. *JSTS J. Semicond. Technol. Sci.* **2001**, *1*, 189–192.
36. Bertaud, T.; Bermond, C.; Blonkowski, S.; Vallee, C.; Lacrevaz, T.; Farcy, A.; Gros-Jean, M.; Flechet, B. Electrical Characterization of Advanced MIM Capacitors With ZrO_2 Insulator for High-Density Packaging and RF Applications. *IEEE Trans. Compon. Packag. Manuf. Technol.* **2012**, *2*, 502–509. [CrossRef]
37. Sharp, K.A.; Honig, B. Electrostatic Interactions in Macromolecules: Theory and Applications. *Annu. Rev. Biophys.* **1990**, *19*, 301–332. [CrossRef]
38. Kittel, C. *Introduction to Solid State Physics*; Wiley: Hoboken, NJ, USA, 2004.
39. Kim, S.J.; Cho, B.J.; Li, M.-F.; Ding, S.-J.; Zhu, C.; Yu, M.B.; Narayanan, B.; Chin, A.; Kwong, D.-L. Improvement of voltage linearity in high-/spl kappa/ MIM capacitors using HfO_2-SiO_2 stacked dielectric. *IEEE Electron Device Lett.* **2004**, *25*, 538–540. [CrossRef]

Disclaimer/Publisher's Note: The statements, opinions and data contained in all publications are solely those of the individual author(s) and contributor(s) and not of MDPI and/or the editor(s). MDPI and/or the editor(s) disclaim responsibility for any injury to people or property resulting from any ideas, methods, instructions or products referred to in the content.

micromachines

Article

3D Light-Direction Sensor Based on Segme Nanorings Combined with Deep Learning

Pengcheng Huang [1], **Peijin Wu** [2], **Ziyuan Guo** [1] **and Zhicheng Ye** [1,3,4,*]

1 Department of Electronic Engineering, Shanghai Jiao Tong I pchhuang@sjtu.edu.cn (P.H.); guo.zy@sjtu.edu.cn (Z.G.)
2 Key Laboratory for Laser Plasmas (Ministry of Education), Tong University, Shanghai 200240, China; wupeijin@sjtu.ed
3 Fujian Science & Technology Innovation Laboratory for Op Fuzhou 350108, China
4 Hangzhou Institute of Optics and Fine Mechanics, Hangzh
* Correspondence: yzhch@sjtu.edu.cn

Abstract: High-precision, ultra-thin angular detectable i promise for light-field detection and reconstruction, th vision and interaction technology. Traditional light-dire nents like gratings and lenses face inherent constraint miniaturization. Recently, angle sensors via coupled dc in attaining high-precision angle perception of incident may herald a novel design paradigm for ultra-compact to measuring the three-dimensional (3D) incident light a sensor concept capable of discerning the 3D light-dire ing structure that is sensitive to both elevation angle (θ) scale and is validated through simulations. Through our simulations reveal that for angle scanning with a detection range of 0∼360° for ϕ and 45°∼90° for can further solve some data aliasing problems to exp the angle sensing dimension based on mutual resona through waveguide implementation or sensor array ar adjusted to accommodate diverse application scenari

Keywords: 3D light-direction sensor; deep learning;

check for updates

Citation: Huang, P.; Wu, P.; Guo, Z.; Ye, Z. 3D Light-Direction Sensor Based on Segmented Concentric Nanorings Combined with Deep Learning. *Micromachines* **2024**, *15*, 1219. https://doi.org/10.3390/mi15101219

Academic Editor: Romeo Bernini

Received: 31 August 2024
Revised: 17 September 2024
Accepted: 18 September 2024
Published: 30 September 2024

Copyright: © 2024 by the authors. Licensee MDPI, Basel, Switzerland. This article is an open access article distributed under the terms and conditions of the Creative Commons Attribution (CC BY) license (https://creativecommons.org/licenses/by/4.0/).

1. Introduction

Information on the incident angle is cruci tion [1]. Relying on the angle information, w image refocusing can thus be implemented [2 implications in machine vision, virtual reality key to designing angle sensors lies in convertir corresponding measurable physical quantitie physical effects with angular sensitivity inclu effect, and the resonance coupling mechanism

Specifically, light projection and masking dence of the projection area on the incident a is the four-quadrant detector (4-QD) [9], whi calculating the proportion of light currents am effect describes the self-imaging property of and the shifts or the depth of self-images ar

by the coordinate angle α in cylindrical coordinates (ρ, α, z) (Figure 1b), whose energy distribution can be modeled as follows [20–25]:

$$i\frac{d}{dt}\binom{a_i(\alpha)}{a_o(\alpha)} = H_0(\theta,\phi)\binom{a_i(\alpha)}{a_o(\alpha)} + H_i(\theta,\phi)\binom{a_i(\alpha)}{a_o(\alpha)} + i\kappa_{f(\alpha,\theta,\phi)}S(f(\alpha,\theta,\phi))\binom{exp(-i\pi sin(f(\alpha,\theta,\phi))d/\lambda)}{exp(i\pi sin(f(\alpha,\theta,\phi))d/\lambda)} \tag{1}$$

where $|a_i(\alpha)|^2$, $|a_o(\alpha)|^2$ represent the stored energy in the inner and outer nanoring cross-sections determined by α, and $f(\alpha,\theta,\phi) = \theta \cdot \cos(\alpha - \phi)$ is the effective angle of incident light received by the cross-sections. H_0 and $H_i(\theta,\phi)$ are the usual Hamiltonian for a pair of resonators and the non-Hermitian Hamiltonian for treating the far-field couple in leakage channels, respectively. The flux $S(f(\alpha,\theta,\phi))$, denoting the incident light energy pumped into the resonators, and the coupling rate $\kappa_{f(\alpha,\theta,\phi)}$ are the main contributors to the angular dependency. The two quantities are determined by the 3D light-direction angles (θ,ϕ), which control the phase $\mp\frac{\mathrm{i}\pi\cos(f(\alpha,\theta,\phi))\mathrm{d}}{\lambda}$ of the excitation wave at each cross-section. Furthermore, the total energy stored in the segments of the inner and outer nanorings determined by the range of $\alpha \in (\alpha_1, \alpha_2)$ can be calculated as follows:

$$e_i = \int_{\alpha_1}^{\alpha_2} |a_i(\alpha)|^2 d\alpha; \;\; e_o = \int_{\alpha_1}^{\alpha_2} |a_o(\alpha)|^2 d\alpha \tag{2}$$

The ratio of the inner and outer energy $r = e_i/e_o$ is a function of (θ,ϕ), from which the 3D light-direction angular sensitivity physical quantity is obtained. For complete concentric nanorings, the response characteristics are polarization-independent due to the strict pairwise nature of the structure, and the energy stored in the inner and outer ring differs from the spacing d and the wavelength of incident light λ, as shown in Figure 2a with the normal incident at s-polarization. In this paper, d is set as 100 nm and λ is set as 550 nm within the absorptive band of Si material to ensure moderate stored energies in both the inner and outer rings for the ratio comparison analysis and actual detection, as shown in Figure 2a. Meanwhile, the (θ,ϕ) influences the resonances within the inner and outer nanorings and further determines the energy stored, as shown at the top and bottom of Figure 2b, respectively.

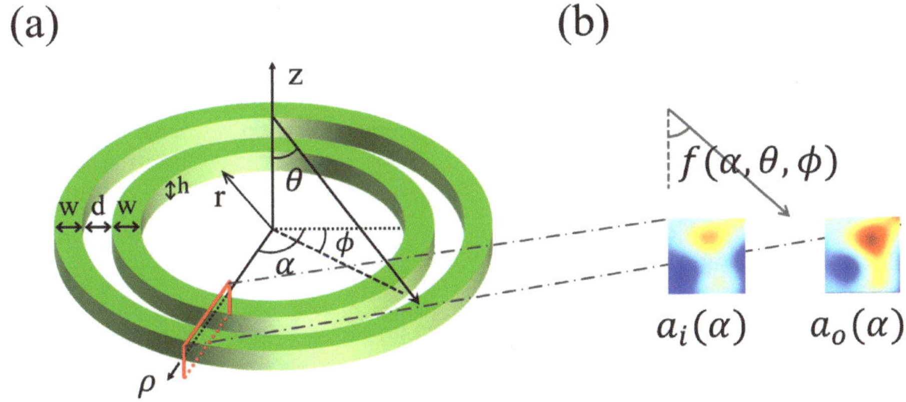

Figure 1. (**a**) Structural diagram of double concentric Si-nanoring resonators. (**b**) For the cross-sections determined by the coordinate angle α in cylindrical coordinates (ρ, α, z), the effective angle of incident light received is a function of (θ, ϕ, α).

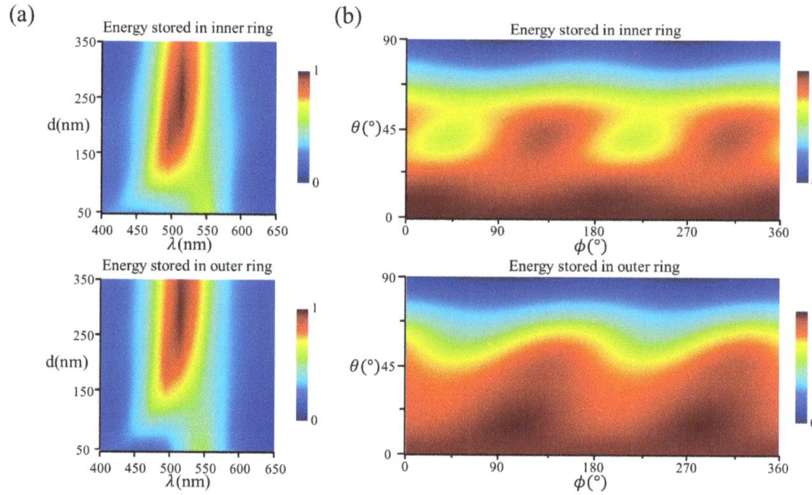

Figure 2. (**a**) Maps of the normalized absorption intensity of the inner and the outer nanorings with spacing (d) and wavelength (λ) under normal incident light; (**b**) numerical maps of the normalized absorption intensity of the inner and the outer nanorings for different (θ, ϕ).

2.2. Three-Dimensional Light-Direction Sensor

To achieve 3D light-direction sensing with a single pair, the response of its structure must exhibit continuous, sensitive, uniform, and monotonous dependence on both the θ and the ϕ simultaneously within the detection range. Due to circular symmetry, the energy stored in the complete nanoring is independent of ϕ. Thus, it is necessary to segment the nanorings to introduce azimuth angle sensitivity, and for full ϕ sensing, at least three segments are needed, as proposed in Figure 3. Assuming that the nanorings are segmented into three pairs—S_1 ($\alpha \in (0, 120°)$), S_2 ($\alpha \in (120°, 240°)$), and S_3 ($\alpha \in (240°, 360°)$)—the energy stored in the inner and outer segments within the same pair can be calculated as follows:

$$e_{i_j/o_j} = \int_{S_j} \left| a_{i_j/o_j}(\alpha) \right|^2 d\alpha,\ j = 1,2,3 \tag{3}$$

where the energy stored in pairs S_1, S_2, S_3 is $S_j = e_{i_j} + e_{o_j}$. According to the equations derived above, physical quantities of energy distribution exhibiting strong angle correlations can be induced, which include the ratio of energy between the inner and the outer segment within the same pair (e_{i_j}/e_{o_j}) and the ratio of energy stored by different pairs $\left(S_j/S_{(j+1) mod\ 3}\right)$. Both quantity ratios demonstrate a strong functional dependence on the angles (θ, ϕ), enabling accurate 3D light-direction angle calculation, which will be elaborated in detail in the discussion section.

In working, the energy in the resonator's segments can be further quantified as photocurrent responses. Thus, the incident angle can then be calculated by the photocurrents exported from the metal (Au/Al) electrodes of the nanoring segments [26]. Additionally, considering the practical working condition and stability of the devices, the structure was fabricated on a silicon dioxide (SiO_2) substrate and most of the architecture, excluding the metal electrodes, was protected by SiO_2 to afford resistance to temperature and humidity.

In summary, the principle of high sensitivity for the segmented concentric nanorings is similar to the angular dependence expounded in the parallel straight nanowires [12]. The non-Hermitian coupling between the inner and outer segments of the same nanoring pair results in a strong angular dependence on elevation. Therefore, θ can be calculated by the ratio of photocurrents between these two segments (RPS). Meanwhile, the differences in responsivities among the nanoring pairs enable azimuthal determination, as the resonance in each pair especially depends on the azimuth. Consequently, ϕ can be calculated by the ratio of total photocurrents between adjacent nanoring pairs (RPPs). Through reasonable adjustments, including the number of segments and structural parameters, higher-precision detection of the 3D light-direction angles can be obtained. Moreover, the proposed structure behaves with high symmetry to ensure translational responses to the polarization of the incident light, indicating computational universality.

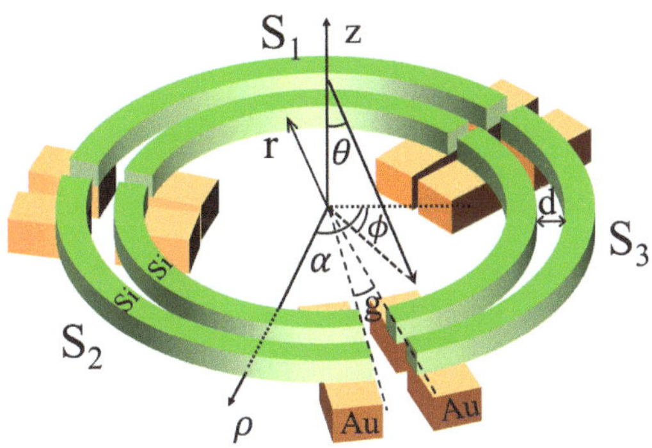

Figure 3. Structural diagram of the 3-segment double-concentric nanoring sensor. Two nanorings are divided into inner and outer rings and further segmented into three pairs: S_1, S_2, and S_3. The inner radius of the nanoring (r) is 1000 nm, the width (w), height (h) of nanorings, and the spacing (d) are 100 nm, and the gap between adjacent nanoring pairs that corresponds to a central angle (g) is 5°.

For the numerical simulations of light energy absorption at different (θ, ϕ) in the proposed segmented concentric nanorings, the 3D finite element method (FEM) by commercial software COMSOL 6.2 was employed in the wavelength domain. As shown in Figure 4, the normalized electric field distribution varies with θ and ϕ, demonstrating obvious angular sensibility due to the resonance in the structure, confirming its functionality for 3D light-direction sensing.

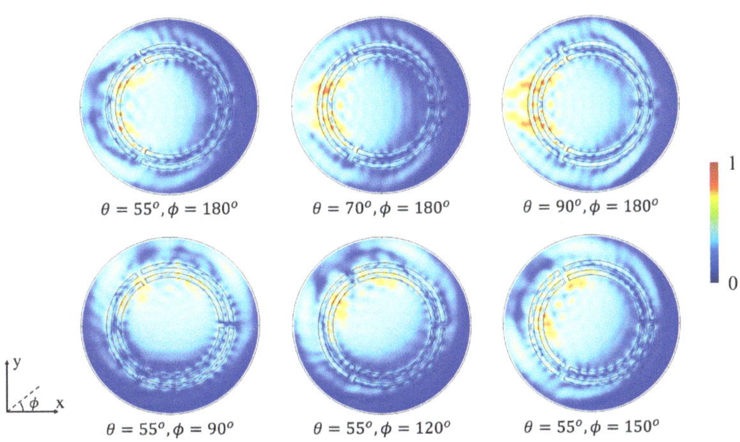

Figure 4. The normalized electric field distribution map for different 3D light-direction angles.

2.3. Deep Learning for Data Fitting

DL offers exceptional capabilities for multidimensional data fitting and inference. According to the universal approximation theorem [27–30], artificial neural networks used for DL can approximate any function under certain conditions. For the proposed angle sensor, modeling the physical relationship between the (θ, ϕ) and several response variables to calculate the angles from these corresponding data is challenging using traditional methods. In detail, the interdependence between θ and ϕ will lead to an expansion of response data dimensions beyond the computing and fitting ability of traditional analysis methods. DL can address this challenge due to its ability to fit high-dimensional data, and it can even potentially enhance both accuracy and the detection range.

To acquire data for DL model training, the proposed sensors are modeled in simulation software and the response photocurrent quantity on each segment is predefined according to Formula (2) as target data. Within the set detection range of θ and ϕ, parameterized scanning with a step size of 1° is deployed to create a dataset, and the data volume is the product of the detection ranges. Furthermore, the dataset is divided into a training set and a testing set with a ratio of 9:1 in mount to jointly optimize the model.

In this work, a fully connected neural network (FCNN) with three hidden layers and a width of 512 neurons per layer was employed, utilizing the rectified linear unit (ReLU) as the activation function. After the sufficient training of 5000 epochs, the DL model could accurately fit and infer θ and ϕ from the response data generated by the sensor structure. The fitting accuracy during the training was measured by the mean squared error (MSE), while the model's performance was quantified by the absolute error (ABS) between the calculated 3D light-direction angles and the targets using a pre-divided test dataset.

2.4. Fabrication Process

The fabrication of the segmented concentric nanorings included electron beam lithography (EBL), deposition, and lift-off, as shown in Figure 5. Specifically, firstly, a 100 nm thick, lightly n-doped poly-Si layer was deposited on the SiO_2 substrate using low-pressure chemical vapor deposition (LPCVD). Then, after the photoresist coating, EBL is employed to define the mask patterns followed by dry etching to create the segmented Si nanorings. Subsequently, a 100 nm thick SiO_2 layer is deposited to cover the nanorings. After lifting off the photoresist, similar processing steps are used to fabricate Au contacts. EBL is used to define the contact patterns, followed by electron-beam physical vapor deposition of Au. The resulting electrodes extend beyond the sensor, facilitating circuit wiring and integration.

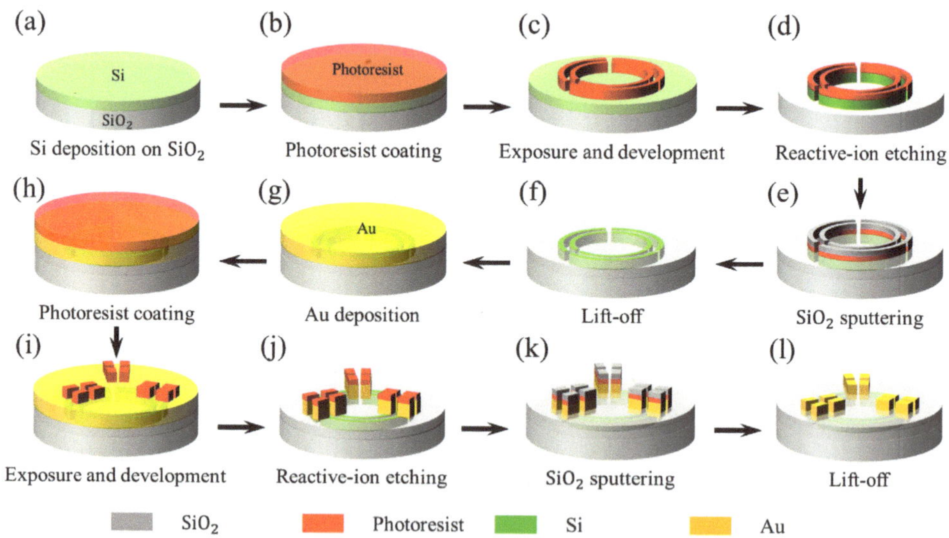

Figure 5. Schematic of the fabrication procedure for the segmented concentric nanorings. (**a**) A 100 nm thick Si on SiO_2 substrate. (**b**) Photoresist coating. (**c**) Electron beam exposure and development. (**d**) Reactive ion etching. (**e**) SiO_2 sputtering. (**f**) Lift-off. (**g**) Au deposition. (**h**) Photoresist coating. (**i**) Electron beam exposure and development. (**j**) Reactive ion etching. (**k**) SiO_2 sputtering. (**l**) Lift-off.

3. Results

In the simulation, the three-segment nanorings shown in Figure 3 represent the minimum segment configuration required to achieve full azimuthal detection. The width (w) and height (h) of the nanorings are both 100 nm, and the spacing (d) between the inner and outer nanorings within the same pair is 100 nm. The inner radius of the nanoring (r) is 1000 nm, and the central angle (g) corresponding to the gap between adjacent nanoring pairs is 5°.

To elucidate the angular dependence of absorption, detailed absorption maps of the sensor under different incident θ and ϕ at the wavelength (λ) of 550 nm are presented in Figure 6. The results clearly show that the normalized total absorption of the inner and outer nanoring segments within the same pair (Figure 6a), the ratio of total absorption between adjacent nanoring pairs (Figure 6b), and the ratio of absorption between the inner and outer nanoring segments within each pair (Figure 6c) all exhibit clear sensitivity to both θ and ϕ within a specific range of θ (Figure 6d).

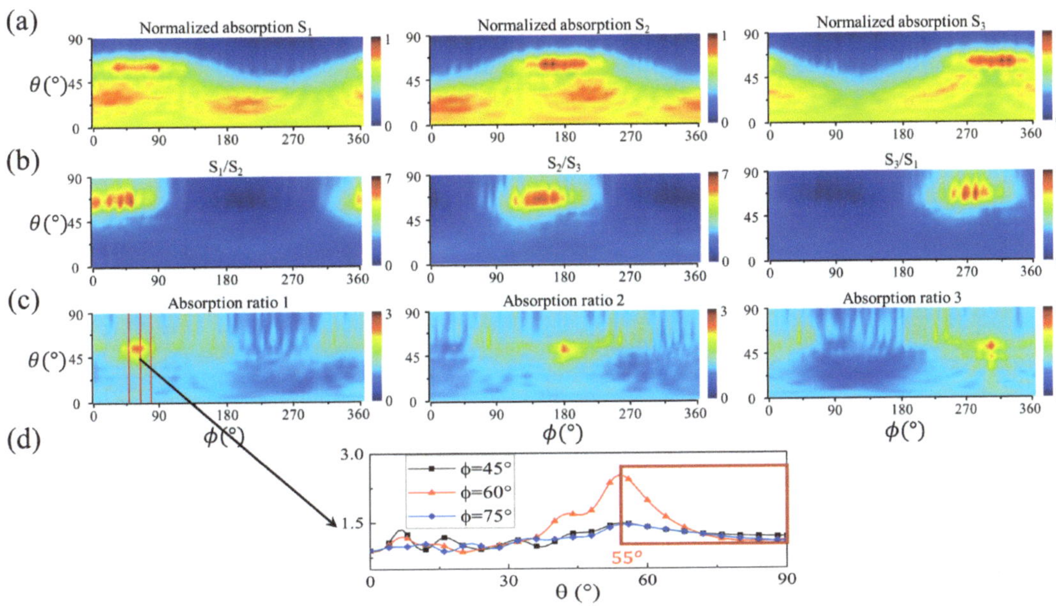

Figure 6. The response of the 3-segment concentric nanoring sensor to the incident light direction angles, θ and ϕ. (**a**) The normalized numerical maps of the total absorption of the S_1, S_2, and S_3 nanoring pairs, including both the inner and outer nanoring segments within each pair. (**b**) The numerical maps of the ratio of the total absorption between adjacent nanoring pairs, S_1/S_2, S_2/S_3, and S_3/S_1. (**c**) The numerical maps of the ratio of the absorption between the inner and outer nanoring segments within each pair, ratio 1, ratio 2, and ratio 3 (e.g., ratio 1 is for the S_1 pair). (**d**) For specific ϕ values, 45°, 60°, and 75°, the absorption ratio for each pair decreases monotonically concerning θ within the range of (55°, 90°), which is the computable range for θ.

From Figure 6d, it can be observed that the detection range for θ, i.e., the monotonic range for a specific ϕ that can be fitted by DL, is approximately 35° (55°∼90°), which was the target range for subsequent calculation. The detection range for θ can be converted and adjusted through a waveguide, where the light in the waveguide obeys the total internal reflection larger than the critical angle. Additionally, due to the translational relationship of responsivities with respect to ϕ between different pairs, the detection range for ϕ is expected to be a full 360°.

Meanwhile, due to the symmetry and translation properties of the proposed structure and its responses as shown, analyzing 1/3 of the sensor is sufficient to verify the overall functionality. We took ϕ within $(60°, 180°)$ as an example, whose calculation method is the same as the other ranges, and θ within the computable range of $(55°, 90°)$, as illustrated earlier. The FCNN is employed to achieve DL, with the training data inputs consisting of S_1/S_2, S_2/S_3, S_3/S_1, ratio 1, ratio 2, ratio 3, and the targets including θ and ϕ.

Based on the test results, the average ABS for θ is 0.372°, with a maximum of 2.077°, and for ϕ, the average ABS is 0.35°, with a maximum of 2.55°. The sensor demonstrates consistent detection capabilities for both θ and ϕ within its detection range, and its performance in (θ, ϕ) calculation is sufficient for most application scenarios.

4. Discussion

4.1. Three-Dimensional Light-Direction Angles Calculation

Based on the angle detection principle of the proposed segmented nanorings, θ can be calculated by the RPS and ϕ can be calculated by the RPP. As shown in Figure 7, the precise (θ, ϕ) can be determined based on the corresponding responses, which exhibit high sensitivities to θ and ϕ.

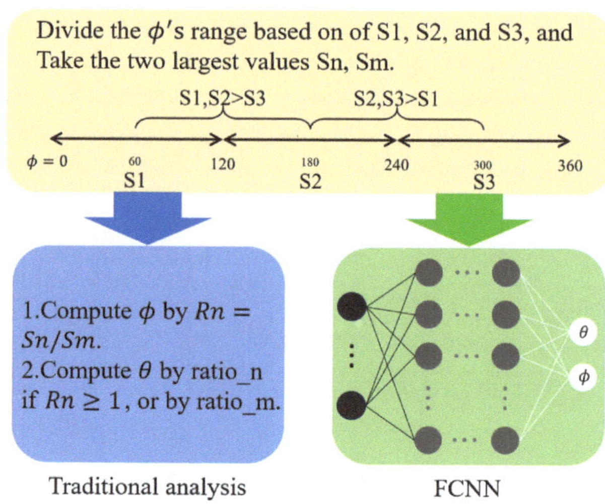

Figure 7. The approach for 3D light-direction angles analysis and calculation. First, an initial division of the azimuth can be made according to the numerical values of the segmented nanoring responses, which helps to select the data sources for accurate (θ, ϕ) calculation. Then, traditional analysis methods or artificial neural networks may be used to compute the θ and ϕ accurately. The former is not feasible because traditional methods are unable to deal with multidimensional data, and we consider using the FCNN to fit the relationship between the responses and the (θ, ϕ), especially the interdependence between the θ and ϕ.

To compute the angles from the responses of the structure, we follow these two steps below: Firstly, we preliminarily divide the range of incident azimuth (ϕ) into 120° steps based on the response amplitudes of the segmented nanoring pairs. This is because the coupling in the pair centered towards the incident light will be the strongest, as shown in Figure 6a. Traditionally, θ might be calculated by RPS and ϕ by RPP independently. However, due to the interdependence between θ and ϕ, this approach is not feasible. To address this issue, we use a FCNN for DL, which treats θ and ϕ as a combined problem, fitting and inferring them from all responses with high precision and a wide-sensing range. This method overcomes the limitations of traditional techniques and leverages the powerful data-fitting capabilities of DL to achieve accurate (θ, ϕ) measurements.

4.2. More Segments to Improve Detection

The number of segments in the proposed sensor significantly impacts the accuracy of the angles' calculation due to its effect on coupling among the nanoring segments. Increasing the number of segments may enhance both the range and accuracy of the (θ, ϕ) detection. Following the same method, four-segment concentric nanorings with the same structural parameters as the three-segment structure were simulated and tested (Figure 8a), with the results partially presented in Figure 8b as examples. Shown as in Figure 8c, the absorption ratio result shows that the detection range for θ was expanded to approximately 45° (45°∼90°), where the absorption ratio descends monotonically with θ for specific ϕ values.

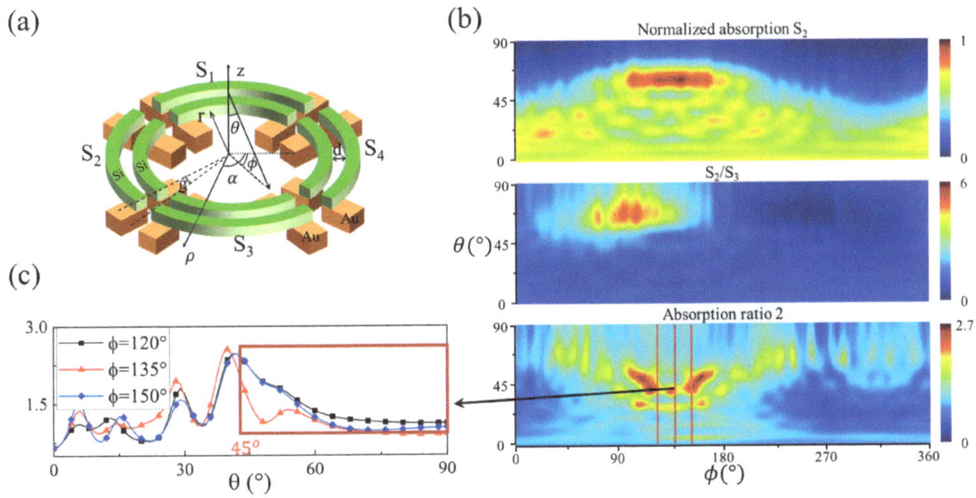

Figure 8. The structural diagram and response of the 4-segment nanorings sensor. (**a**) The structure of the 4 segments. The nanorings are segmented into four pairs—S_1, S_2, S_3, and S_4—with the same structural parameters as the 3-segment structure. In detail, the inner radius of the nanoring (r) is 1000 nm, the width (w) and height (h) of nanorings and the spacing (d) are 100 nm, and the gap between adjacent nanoring pairs corresponds to a central angle (g) of 5°. (**b**) The response of the 4-segment structure. As with the 3 segments, the responses of the 4 segments also show translational symmetry. We present a normalized numerical map of the total absorption of the S_2 pair, the numerical map of the ratio of S_2 and S_3, and the corresponding absorption ratio of the inner and outer rings within S_2 (ratio 2) as examples. (**c**) From the characteristics of ratio 2, it can be seen that the monotonical range of θ has expanded to approximately (45°, 90°), which is the computable range for θ.

As with the 3-segment structure, a DL network is used to fit the response data and determine the 3D light-direction angles for the 4-segment configuration. As shown in the results, the average ABS for θ is 0.187°, with a maximum of 2.252°, and for ϕ, the average ABS is 0.174°, with a maximum of 1.704°, indicating that the angular detection capability has been further improved to be better than that of the 3 segments due to more segments being involved. However, it is important to note that although increasing the number of segments may improve the detection range and accuracy, it also increases manufacturing difficulty and costs; thus, a trade-off needs to be considered.

4.3. Cross-Arranged Nanowire Pairs for 3D Light-Direction Sensing

Additionally, the combination of resonant nanowires can be intuitively expected to expand the detection dimension and function of an angle sensor (Figure 9a). In the simulation, nanowires with a width and height of 100 nm and a length of 500 nm, wrapped in SiO_2, were divided into two pairs that were internally coupled and arranged in a vertical cross configuration with a specified distance (gap) between them.

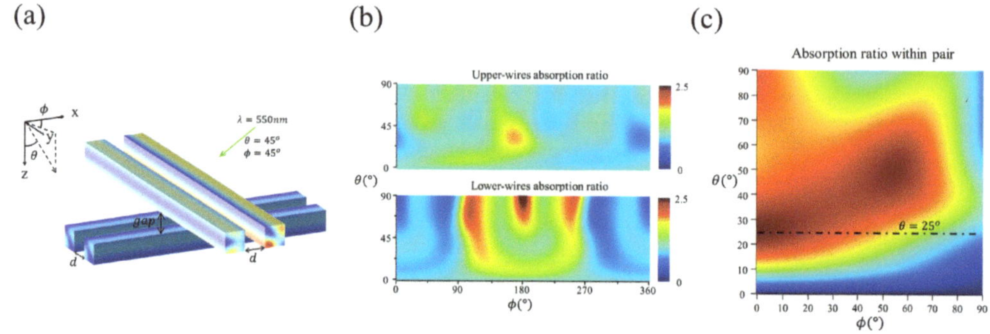

Figure 9. The structure and response of the cross-arranged nanowires sensor to the (θ, ϕ). (**a**) Two pairs of vertically cross-arranged coupled nanowires, with a vertical distance (gap) of 100 nm. The nanowires have a width and height of 100 nm, a length of 500 nm, and an internal spacing (d) of 100 nm within each nanowire pair. (**b**) The numerical maps of the absorption ratio within the upper and lower nanowire pairs with a gap of 100 nm concerning θ and ϕ. The response of the upper layer is different from that of the lower layer, and they do not exhibit a consistent regularity over a wide range of angles when computing the angles. (**c**) The numerical map of the absorption ratio between the two nanowires within a pair when the gap is 1 μm. When the gap is large enough ($gap \gg \lambda$), the two pairs can be analyzed separately, and their responses have angular symmetry. According to the map, the absorption ratio within a pair is monotonically changing concerning θ for a specific ϕ when $\theta \in (0, 25^\circ)$, and this range is the detection range for θ.

With a gap of 100 nm, as shown in Figure 9b, the response of the composite structure for different θ and ϕ does not exhibit a consistent regularity over a wide range, making it impossible to accurately reconstruct large-range 3D light-direction angles. The response of the upper layer differs from that of the lower layer, which is likely due to the interference of the upper-layer nanowires with the incident light, causing the light received by the lower-layer nanowires to differ from the original incident light.

To eliminate the interference between the upper layer and the lower layer, when the gap is large enough ($gap \gg \lambda$), taking a gap of 1 μm as an example, the response of the two pairs becomes symmetrical. The absorption ratio between the two nanowires within a pair is shown in Figure 9c, exhibiting regular characteristics within a certain range of $\theta \in (0, 25^\circ)$. In this range, the RPN changes monotonically with respect to θ for a specific ϕ, enabling the calculation of the (θ, ϕ) via DL.

Using the same DL fitting method as before for $\theta \in (0, 25^\circ)$, and taking $\phi \in (0, 45^\circ)$ as an example that represents the entire structure because of its symmetry, the average ABS for θ is 0.198°, with a maximum of 0.578°, and that for ϕ is 0.948°, with a maximum of 16.387° in the test. It is clear that although the cross-arranged nanowire-pairs can detect (θ, ϕ) when the distance between the upper layer and the lower layer is sufficiently large, the calculation error for ϕ is significantly higher than for θ, which indicates that the sensitivity of the device response to these two angles is inconsistent, rendering it impractical for applications requiring uniform sensitivity.

By contrast, the proposed segmented nanorings, with their segmented and circular characteristics, can sense 3D light-direction angles with a smaller thickness, a wider detection range, and uniform angular sensitivity (Table 1), which demonstrates that the segmented nanoring structure offers better and more stable angle detection capabilities and a wider detection range than the cross-arranged nanowire pairs. The configuration of the segmented concentric nanorings ensures consistent detection capabilities for both θ and ϕ, making it more suitable for practical applications, and it can be inferred that more segments will lead to further improvement.

Table 1. The absolute error of θ and ϕ for different angular detection structures, including the detection range for θ (the detection ranges for ϕ are all 360°), average absolute error, maximum absolute error, and the variance.

Structure	Detection Range for θ (°)	Absolute Error of θ (°)			Absolute Error of ϕ (°)		
		Average	Maximum	Variance	Average	Maximum	Variance
3-segment nanorings	55–90	0.372	2.077	0.121	0.350	2.550	0.102
4-segment nanorings	45–90	0.187	2.252	0.041	0.174	1.704	0.030
Cross-arranged nanowire pairs	0–25	0.198	0.578	0.009	0.948	16.387	7.291

The FCNNs deployed for these three sensor structures have the same configuration and training parameters, except for the different numbers of input layer neurons due to the varying response amounts of different structures (6, 8, 4, respectively). In detail, the FCNNs consist of three hidden layers with a width of 512 and one output layer with a width of 2 (512 × 512 × 512 × 2), and they are trained for a total of 5000 epochs based on an Adam optimizer, with a learning rate of 10^{-4} using the ReLU as the activation function.

As shown in Figure 10, compared to the cross-arranged nanowire pairs structure, the segmented nanorings have smaller convergence errors, and dividing into more segments can result in faster convergence speeds and smaller losses during the DL training process. These improvements may stem from the expansion of response data dimensions brought about by additional segments and the stronger resonance caused by smaller nanoring lengths, which provide computational benefits such as cross-validation between these data in the DL model.

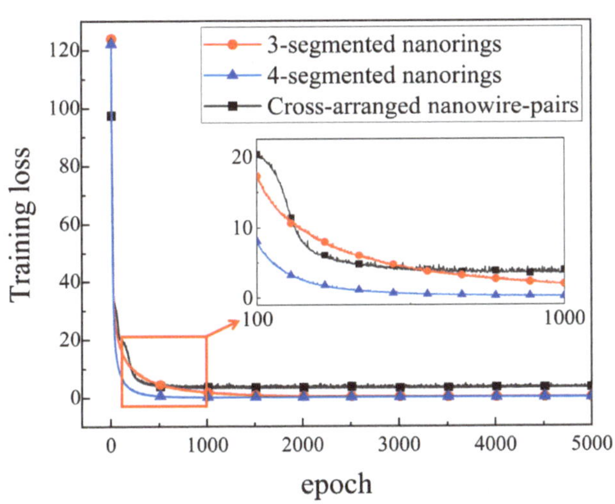

Figure 10. The loss descent curve during the training process of deep learning models corresponding to different structures. (To distinguish clearly, the first 10 epochs corresponding to the loss have been taken out).

5. Conclusions

To achieve precise 3D light-direction angle detection with a single sensing unit, we propose a structure of segmented concentric nanoring pairs absorptive resonators as a possible sensor design approach. In the simulations, this structure exhibits strong and uniform angular sensitivity to both elevation and azimuth angles. By employing deep learning for data fitting, the sensor can achieve a detection range of $0 \sim 360°$ for azimuth and $45° \sim 90°$ for elevation, with an average accuracy of 0.19° in (θ, ϕ) angle sensing, thereby meeting the requirements of most application scenarios. The accuracy and range of angle sensing can be improved with an increase in the number of segments. Finally, based on the structure of segmented concentric nanorings, by combining the resonance characters with deep learning, precise angle sensing is possible, showing various promising applications such as in optical field reconstruction, dynamic ranging, and optical interaction, and even military reconnaissance and guidance could be achieved by array integration. It should be pointed out that the proposed sensor was only validated through simulations, but we believe that through further exploration and experimentation, practical applications will be inspired to come to fruition in the near future.

Author Contributions: Conceptualization, P.H., P.W. and Z.G.; methodology, P.H., P.W. and Z.Y.; software, P.H. and P.W.; validation, P.H.; investigation, P.H.; data curation, P.H. and Z.G.; Funding acquisition, Z.Y.; writing—original draft preparation, P.H. and P.W.; writing—review and editing, P.H. and Z.Y. All authors have read and agreed to the published version of the manuscript.

Funding: This research was funded by the Natural Science Foundation of the Fujian Province, China, grant number 2023J01234.

Data Availability Statement: The raw data supporting the conclusions of this article will be made available by the authors on request.

Conflicts of Interest: The authors declare no conflicts of interest.

References

1. Adelson, E.H.; Bergen, J.R. *The Plenoptic Function and the Elements of Early Vision*; Vision and Modeling Group, Media Laboratory, Massachusetts Institute of Technology: Cambridge, MA, USA, 1991; Volume 2.
2. Ng, R.; Levoy, M.; Brédif, M.; Duval, G.; Horowitz, M.; Hanrahan, P. Light Field Photography with a Hand-Held Plenoptic Camera. Ph.D. Thesis, Stanford University, Stanford, CA, USA, 2005.
3. Wang, A.; Gill, P.; Molnar, A. Light field image sensors based on the Talbot effect. *Appl. Opt.* **2009**, *48*, 5897–5905. [CrossRef] [PubMed]
4. Wang, A.; Molnar, A. A light-field image sensor in 180 nm CMOS. *IEEE J. Solid-State Circuits* **2011**, *47*, 257–271. [CrossRef]
5. Li, Q.; van de Groep, J.; Wang, Y.; Kik, P.G.; Brongersma, M.L. Transparent multispectral photodetectors mimicking the human visual system. *Nat. Commun.* **2019**, *10*, 4982. [CrossRef] [PubMed]
6. Yu, L.; Feng, X.; Hu, P.; Lin, X.; Jing, T. Development of a 3-DOF Angle Sensor Based on a Single Laser Interference Probe. *Micromachines* **2023**, *14*, 2221. [CrossRef]
7. Koch, C.; Oehm, J.; Emde, J.; Budde, W. Light source position measurement technique applicable in SOI technology. *IEEE J. Solid-State Circuits* **2008**, *43*, 1588–1593. [CrossRef]
8. Yi, S.; Xiang, J.; Zhou, M.; Wu, Z.; Yang, L.; Yu, Z. Angle-based wavefront sensing enabled by the near fields of flat optics. *Nat. Commun.* **2021**, *12*, 6002. [CrossRef]
9. Li, D.; Liu, S. Research on four-quadrant detector and its precise detection. *Int. J. Digit. Content Technol. Appl.* **2011**, *5*, 138–143.
10. Azaña, J.; Guillet de Chatellus, H. Angular talbot effect. *Phys. Rev. Lett.* **2014**, *112*, 213902. [CrossRef]
11. Born, M. *Principles of Optics-Electromagnetic Theory of Propagation, Interference and Diffraction of Light*, 7th ed.; Elsevier: Amsterdam, The Netherlands, 1999.
12. Yi, S.; Zhou, M.; Yu, Z.; Fan, P.; Behdad, N.; Lin, D.; Wang, K.X.; Fan, S.; Brongersma, M. Subwavelength angle-sensing photodetectors inspired by directional hearing in small animals. *Nat. Nanotechnol.* **2018**, *13*, 1143–1147. [CrossRef]
13. Nagarajan, A.; Hara, S.; Satoh, H.; Panchanathan, A.P.; Inokawa, H. Angle-sensitive detector based on silicon-on-insulator photodiode stacked with surface plasmon antenna. *Sensors* **2020**, *20*, 5543. [CrossRef]
14. Barlow, H.B. The size of ommatidia in apposition eyes. *J. Exp. Biol.* **1952**, *29*, 667–674. [CrossRef]
15. Zhu, X.; Liang, B.; Kan, W.; Peng, Y.; Cheng, J. Deep-subwavelength-scale directional sensing based on highly localized dipolar mie resonances. *Phys. Rev. Appl.* **2016**, *5*, 054015. [CrossRef]
16. Lee, T.; Nomura, T.; Su, X.; Iizuka, H. Fano-Like Acoustic Resonance for Subwavelength Directional Sensing: 0–360 Degree Measurement. *Adv. Sci.* **2020**, *7*, 1903101. [CrossRef]

17. Meng, Y.; Hu, X.; Yang, C.; Shen, X.; Cao, X.; Lin, L.; Yan, F.; Yue, T. Angle-sensitive Pixels Based on Subwavelength Compound Gratings. *Curr. Opt. Photonics* **2022**, *6*, 359–366.
18. Yi, L.; Hou, B.; Zhao, H.; Liu, X. X-ray-to-visible light-field detection through pixelated colour conversion. *Nature* **2023**, *618*, 281–286. [CrossRef]
19. Bakirova, L.I.; Voronkov, G.S.; Lyubopytov, V.S.; Butt, M.A.; Khonina, S.N.; Stepanov, I.V.; Grakhova, E.P.; Kutluyarov, R.V. Micro-Ring Resonator-Based Tunable Vortex Beam Emitter. *Micromachines* **2023**, *15*, 34. [CrossRef] [PubMed]
20. Suh, W.; Wang, Z.; Fan, S. Temporal coupled-mode theory and the presence of non-orthogonal modes in lossless multimode cavities. *IEEE J. Quantum Electron.* **2004**, *40*, 1511–1518.
21. Torres, J.; Baptista, A.; Machado, V.M. Coupling analysis in concentric ring waveguides. *J. Light. Technol.* **2013**, *31*, 2140–2145. [CrossRef]
22. Verslegers, L.; Yu, Z.; Ruan, Z.; Catrysse, P.B.; Fan, S. From Electromagnetically Induced Transparency to Superscattering with a Single Structure: A Coupled-Mode Theory for Doubly Resonant Structures. *Phys. Rev. Lett.* **2012**, *108*, 083902. [CrossRef]
23. Verslegers, L.; Yu, Z.; Catrysse, P.B.; Fan, S. Temporal coupled-mode theory for resonant apertures. *JOSA B* **2010**, *27*, 1947–1956. [CrossRef]
24. Zhang, S.; Ye, Z.; Wang, Y.; Park, Y.; Bartal, G.; Mrejen, M.; Yin, X.; Zhang, X. Anti-Hermitian Plasmon Coupling of an Array of Gold Thin-Film Antennas for Controlling Light at the Nanoscale. *Phys. Rev. Lett.* **2012**, *109*, 193902. [CrossRef] [PubMed]
25. Zhou, M.; Shi, L.; Zi, J.; Yu, Z. Extraordinarily large optical cross section for localized single nanoresonator. *Phys. Rev. Lett.* **2015**, *115*, 023903. [CrossRef] [PubMed]
26. Cao, L.; White, J.S.; Park, J.-S.; Schuller, J.A.; Clemens, B.M.; Brongersma, M.L. Engineering light absorption in semiconductor nanowire devices. *Nat. Mater.* **2009**, *8*, 643–647. [CrossRef] [PubMed]
27. Cybenko, G. Approximation by superpositions of a sigmoidal function. *Math. Control Signals Syst.* **1989**, *2*, 303–314. [CrossRef]
28. Hornik, K.; Stinchcombe, M.; White, H. Multilayer feedforward networks are universal approximators. *Neural Netw.* **1989**, *2*, 359–366. [CrossRef]
29. Leshno, M.; Lin, V.Y.; Pinkus, A.; Schocken, S. Multilayer feedforward networks with a nonpolynomial activation function can approximate any function. *Neural Netw.* **1993**, *6*, 861–867. [CrossRef]
30. Chen, T.; Chen, H. Universal approximation to nonlinear operators by neural networks with arbitrary activation functions and its application to dynamical systems. *IEEE Trans. Neural Netw.* **1995**, *6*, 911–917. [CrossRef]

Disclaimer/Publisher's Note: The statements, opinions and data contained in all publications are solely those of the individual author(s) and contributor(s) and not of MDPI and/or the editor(s). MDPI and/or the editor(s) disclaim responsibility for any injury to people or property resulting from any ideas, methods, instructions or products referred to in the content.

micromachines

MDPI

Article

Fabrication of Electrospun Porous TiO_2 Dielectric Film in a Ti–TiO_2–Si Heterostructure for Metal–Insulator–Semiconductor Capacitors

Jin-Uk Yoo †, Tae-Min Choi † and Sung-Gyu Pyo *

School of Integrative Engineering, Chung-Ang University, 84, Heukseok-ro, Dongjak-gu, Seoul 06974, Republic of Korea; wlsdnr5771@naver.com (J.-U.Y.); c79411@gmail.com (T.-M.C.)
* Correspondence: sgpyo@cau.ac.kr
† These authors contributed equally to this work.

Abstract: The development of metal–insulator–semiconductor (MIS) capacitors requires device miniaturization and excellent electrical properties. Traditional SiO_2 gate dielectrics have reached their physical limits. In this context, high-k materials such as TiO_2 are emerging as promising alternatives to SiO_2. However, the deposition of dielectric layers in MIS capacitors typically requires high-vacuum equipment and challenging processing conditions. Therefore, in this study, we present a new method to effectively fabricate a poly(vinylidene fluoride) (PVDF)-based TiO_2 dielectric layer via electrospinning. Nano-microscale layers were formed via electrospinning by applying a high voltage to a polymer solution, and electrical properties were analyzed as a function of the TiO_2 crystalline phase and residual amount of PVDF at different annealing temperatures. Improved electrical properties were observed with increasing TiO_2 anatase content, and the residual amount of PVDF decreased with increasing annealing temperature. The sample annealed at 600 °C showed a lower leakage current than those annealed at 300 and 450 °C, with a leakage current density of 7.5×10^{-13} A/cm^2 when Vg = 0 V. Thus, electrospinning-based coating is a cost-effective method to fabricate dielectric thin films. Further studies will show that it is flexible and dielectric tunable, thus contributing to improve the performance of next-generation electronic devices.

Keywords: MIS; electrospinning; TiO_2; leakage current; electrospun dielectric layer

check for updates

Citation: Yoo, J.-U.; Choi, T.-M.; Pyo, S.-G. Fabrication of Electrospun Porous TiO_2 Dielectric Film in a Ti–TiO_2–Si Heterostructure for Metal–Insulator–Semiconductor Capacitors. *Micromachines* **2024**, *15*, 1231. https://doi.org/10.3390/mi15101231

Academic Editors: Sadia Ameen and Ai-Qun Liu

Received: 11 September 2024
Revised: 27 September 2024
Accepted: 30 September 2024
Published: 30 September 2024

Copyright: © 2024 by the authors. Licensee MDPI, Basel, Switzerland. This article is an open access article distributed under the terms and conditions of the Creative Commons Attribution (CC BY) license (https://creativecommons.org/licenses/by/4.0/).

1. Introduction

In recent decades, the advancement of memory device structures, particularly in gate and capacitor dielectrics, has driven significant innovation. As device dimensions continue to scale down, high-k dielectric materials, such as titanium dioxide (TiO_2) [1,2], hafnium dioxide (HfO_2) [3], and zirconium dioxide (ZrO_2) [4], have been increasingly integrated into cutting-edge technologies like FinFETs and high-k metal gate (HKMG) structures to mitigate leakage current and short-channel effects [5].

Among these materials, TiO_2 stands out due to its exceptionally high dielectric constant, ranging between 80 and 120, depending on whether it crystallizes in the anatase or rutile phase [6]. However, despite its high dielectric constant, TiO_2 struggles with controlling leakage currents, primarily due to its relatively narrow band offset. To enhance its electrical properties, reducing oxygen vacancies and improving structural order through annealing processes are critical, as the phase transformation from anatase to rutile significantly impacts both its dielectric behavior and leakage characteristics [7–10].

While these high-k materials deliver superior performance, they have deposition methods such as physical vapor deposition (PVD) [11], chemical vapor deposition (CVD) [12], plasma-enhanced CVD (PECVD) [13], and atomic layer deposition (ALD) [10,14–16]. However, these deposition methods involve long processing times and high costs, owing to the use of ultrahigh vacuum systems, and they are limited by equipment space requirements [17]. Therefore, it is crucial to identify novel techniques that can significantly reduce the deposition time and cost, while also enabling deposition at the nano- and microscale under room temperature and atmospheric pressure conditions.

In the electrospinning process, nanoscale fibers are fabricated by applying a high voltage to a polymer solution [18,19]. In general, electrospun one-dimensional fibers are formed when the solution concentration is above the critical point; however, when the concentration is below the critical point, a porous film ranging from tens of nanometers to hundreds of micrometers can be formed by changing from electrospinning to electrospraying. In addition, depending on the metal nozzle type, it is possible to produce fibers with hollow, core–shell, and porous structures. Also, the deposition method using electrospinning does not use vacuum equipment and the equipment itself is inexpensive, so the process cost difference is about 5–10 times compared to the conventional method. The advantages of electrospinning are not the only ones. The film adhesion can be adjusted according to the type of organic material, and the dielectric constant can be controlled by tuning the doping material; hence, this approach can be developed into a coating technology with unlimited potential [20–24].

In this paper, an anatase TiO_2 dielectric layer was deposited on a p-type (100)-oriented bare wafer using a simple, fast, and low-cost PVDF-based electrospinning method. To investigate the dependence of the MIS capacitor performance on the residual PVDF and the TiO_2 phase, the annealing process was optimized at rapid thermal annealing (RTA) temperatures of 300, 450, and 600 °C. Then, a Ti metal layer was deposited by sputtering using a Ti target to fabricate an anatase TiO_2-based heterostructured MIS capacitor. The electrospinning method provides a new way to tailor the electrical properties of MIS capacitors by controlling various parameters, in order to form films with the desired structure according to the target application. This paper highlights that electrospinning can be performed without vacuum equipment, making the process very low cost and applicable to semiconductor applications. However, further research is required for nanoscale processing of semiconductors.

2. Experimental Section

First, the solution was prepared to fabricate the TiO_2 dielectric layer, followed by the electrospinning process

2.1. Materials and Solution Preparation

Titanium(IV) oxide (TiO_2, anatase form, JUNSEI, Tokyo, Japan) and PVDF (Mw~534,000, Sigma Aldrich, Seoul, Republic of Korea), acting as binder, were mixed in a 94:6 (4.7 g of TiO_2, 0.3 g of PVDF) mass ratio; then, N-methyl-2-pyrrolidone (NMP, 99.7%, JKC, Cheonan, Republic of Korea), used as solvent, was added at about 7 mL to obtain the appropriate viscosity. The solution was stirred at 800 rpm for 24 h (magnetic stirrer, WISD, Seoul, Republic of Korea). Additionally, sonication and stirring were repeated several times in an ultrasonic cleaner (DAIHAN Scientific, Seoul, Republic of Korea) to disperse the TiO_2 phase.

2.2. Electrospinning

In general, the electrospinning process is used to produce nano–microscale fibers; however, in order to prepare a relatively uniform film and increase the adhesion between the silicon wafer and the dielectric layer, PVDF-based electrospraying was performed with the concentration of the electrospinning solution reduced to ~3–4 wt.%. In this experiment, PVDF was used for the electrospinning polymer solution. PVDF has various advantages such as high production stability and abundant green resources, and the PVDF phase can be controlled by electrospinning, which makes it a promising electrode material. PVDF is also used as an anode binder in lithium-ion batteries and is often used to improve the structural stability, so it was also used in this experiment [25–27]. Electrospinning was performed using a single nozzle (nozzle adaptor, NanoNC, Seoul, Republic of Korea) and a constant flow rate of 2 mL/h using a syringe pump (Fusion 100-X precision dosing two-channel syringe pump, Chemyx, Stafford, TX, USA). To improve the uniformity of the dielectric layer, a drum-type collector (NNC-DC90H, NanoNC, Seoul, Republic of Korea) was used as a spin coater, and the wafer was spun in place at 200 rpm. The distance between the drum collector and the syringe tip was set to 17 cm, and a high voltage of 8–10 kV was applied. Finally, a 23-gauge plastic nozzle (inner diameter: 0.33 mm, outer diameter: 0.63 mm, NanoNC, Seoul, Republic of Korea) was used to prevent clogging at the tip when TiO_2 particles flowed through the syringe at a constant rate.

2.3. Fabrication of MIS Capacitor

In this experiment, rapid thermal process (RTP) annealing (KVR-2000, Korea Vacuum Tech, Goyang, Republic of Korea) was used to anneal the TiO_2 dielectric layer at RTA temperatures of 300, 450, and 600 °C to observe changes in crystallinity and electrical properties with the annealing temperature. In the RTP, the N_2 gas was flowed at 500 sccm under vacuum conditions. Afterward, the sputtering device (RF magnetron sputtering system KVS-2004, Korea Vacuum Tech) was used to deposit the metal layer. Using a Ti target, the Ar gas was flowed inside the chamber at 18 sccm under high vacuum conditions. After adjusting the pressure inside the chamber to 2.2×10^{-2} torr, the Ar plasma was formed with an RF power of 200 mW to deposit Ti for 120 min; finally, the MIS Ti–TiO_2–Si heterostructure was successfully fabricated (Figure 1).

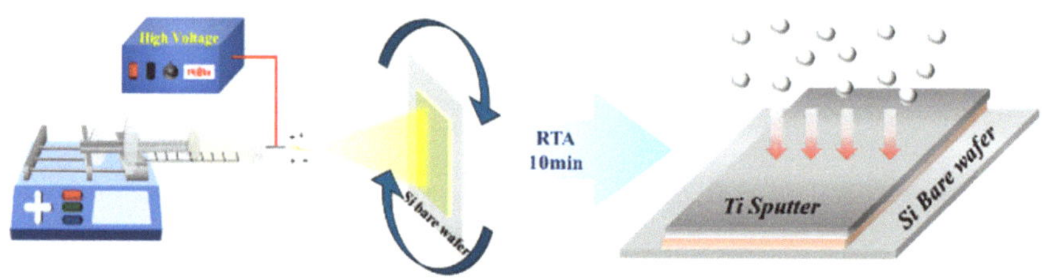

Figure 1. Schematic illustration of the fabrication of the electrospun TiO_2 dielectric layer and Ti-sputtered metal layer.

2.4. Analysis

Scanning electron microscopy (SEM), atomic force microscopy (AFM), and Raman spectroscopy were used to evaluate the surface state, roughness, adhesion force, and composition of the electrospun MIS capacitor (see Table S1). X-ray diffraction (XRD) analysis was performed to determine crystalline phase changes with the annealing temperature. Finally, a probe station was used to evaluate the electrical properties.

Raman spectroscopy measurements employed a XperRAM35V (NanoBase, Seoul, Republic of Korea) instrument with a 532 nm laser and a 1800-lpmm grating. AFM experiments were carried out with a NX-10 (ParkSystems, Suwon, Republic of Korea) instrument; a Tap300Al-G cantilever (force constant 40 N/m, resonance frequency 300 kHz) was applied for non-contact mode experiments, whereas a PPP-CONTSCR cantilever (force constant 40 N/m, resonance frequency 300 kHz) was used for measuring force–distance curves. Field-emission SEM (FE-SEM, SIGMA300, Carl Zeiss, Oberkochen, Germany) was used to analyse the deposition conditions, thickness, and interlayer separation of the film. Finally, the crystal structure of TiO_2 was characterized by XRD using Cu Kα radiation (New D8-Advance, Bruker-AXS, Karlsruhe, Germany).

3. Results and Discussion

Figure 2 shows SEM images of electrospun TiO_2 samples annealed at RTA temperatures of 300, 450, and 600 °C. The above samples were electrospun at 2 mL/h for a total of 5 min, and the dielectric layer had a thickness range of ~6–10 μm. A thinner film could be produced by reducing the electrospinning process time. Figure 2a shows the final electrospun MIS structure obtained in this study. Figure 2a is a cross-sectional SEM image of a hetero-structured MIS capacitor with a TiO_2 dielectric layer deposited by electrospinning and a Ti metal layer deposited by sputtering after 600 °C RTA heat treatment. A Ti metal layer was deposited on the electrospun TiO_2 using a sputter, as shown in more detail in Figure S1. Figure 2b shows the electrospun MIS structure of the sample annealed at an RTA temperature of 300 °C for 10 min. A large amount of residual PVDF can be seen in the figure. Moreover, Figure 2c shows the sample annealed at an RTA temperature of 450 °C for 10 min; a small amount of PVDF was still present, although much smaller than that of the 300 °C-annealed sample. Finally, Figure 2d shows the sample annealed at an RTA temperature of 600 °C for 10 min: most PVDF was volatilized during the annealing process, and no residual PVDF was visible under the optics. The amount of PVDF in the dielectric layer affects the leakage current. In this experiment, PVDF was used to act as a binder for TiO_2. Although PVDF can improve the insulation properties, it exhibits polarization due to the hysteresis effect in the electric field, which leads to energy loss [28]. Therefore, the amount of residual PVDF will affect the leakage current. To confirm the morphology and roughness of electrospun TiO_2 as a function of temperature, the AFM data are shown in detail in Figure S2.

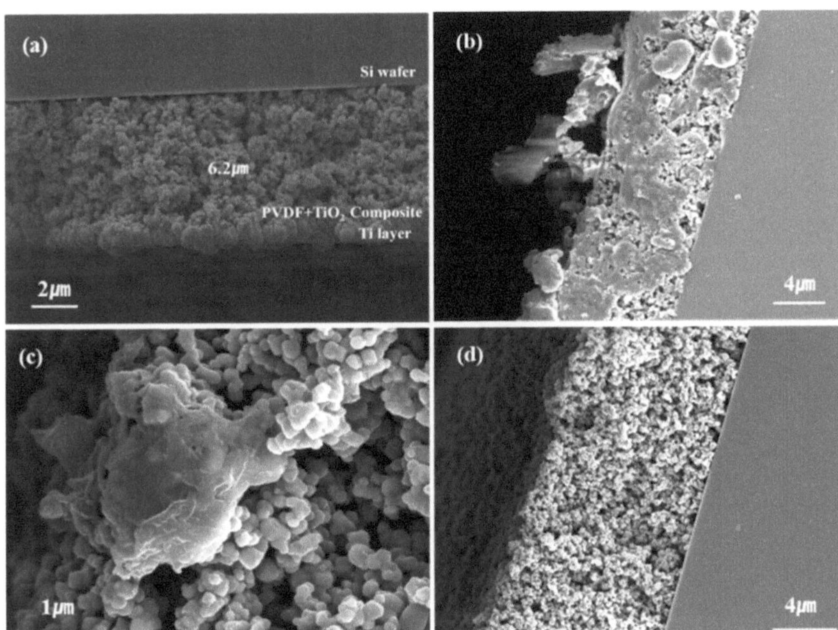

Figure 2. Cross-section of FE-SEM images of (**a**) electrospun porous TiO_2 dielectric films in the Ti–TiO_2–Si heterostructure for MIS capacitors annealed at 600 °C, as well as electrospun TiO_2 dielectric films annealed at RTA temperatures of (**b**) 300 °C, (**c**) 450 °C, and (**a**,**d**) 600 °C.

The XRD patterns in Figure 3 show that the crystallinity of the TiO_2 thin films improved with increasing annealing temperature. The XRD analysis was performed before the Ti electrode was deposited. TiO_2 anatase phase peaks corresponding to (101), (103), (004), (112), (200), (105), and (211) orientations were observed at $2\theta = 25.7°$, 37.1°, 38°, 38.7°, 48°, 54°, and 55°, respectively [29]. More details of XRD patterns of the TiO_2 electrospun thin films are shown in Figure S3. The intensity of the (101) peak, representative of anatase TiO_2, increased with the RTA temperature. The anatase peak intensities of the 450 and 600 °C annealed samples increased by ~9.3% and 25.2%, respectively, compared to that of the sample annealed at 300 °C. Kang et al. reported that, among the two main TiO_2 phases, rutile showed a worse leakage current than anatase. It was also reported that the leakage current decreased as the proportion of anatase phase increased, denoting better electrical properties [30]. Therefore, the anatase peak ratio was expected to increase as the annealing temperature increases from 300 to 450 and 600 °C, reflecting better leakage current properties. As shown in Figure 3b, the peaks shifted toward lower 2θ values as the annealing temperature increased. This was because, when the material was annealed, the atoms underwent thermal expansion; this caused the lattice constant to increase, with a corresponding shift of the XRD peaks toward lower angles [31].

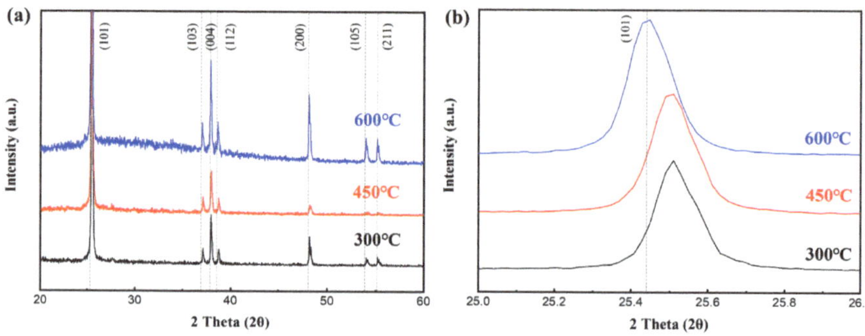

Figure 3. XRD patterns of electrospun TiO_2 dielectric layers annealed at RTA temperatures of 300, 450, and 600 °C; 2θ ranges: (**a**) 20°–65°, (**b**) 25°–26°.

Raman spectroscopy can be used to qualitatively evaluate the percentage of residual PVDF in the TiO_2/PVDF composite thin film and the crystallinity of TiO_2 particles. As shown in Figures 4 and S4, TiO_2 anatase peaks (144, 394, 514, 634 cm^{-1}) were observed for all samples at 300, 450, and 600 °C, and the peak intensity tended to increase as the annealing temperature increased. In addition, the fluorescence caused by PVDF tended to decrease with increasing annealing temperature. This matches the trend of the SEM images, which indicates that the material composition ratio and thin film properties of the TiO_2/PVDF composite can be controlled by adjusting the annealing temperature conditions.

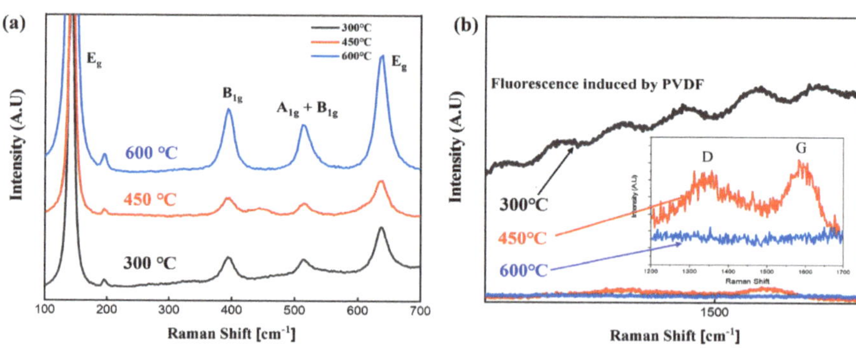

Figure 4. Raman data of electrospun TiO_2 dielectric layer annealed at RTA temperatures of 300, 450, and 600°. (**a**) Comparison of TiO_2 peaks; (**b**) enlargement of the organic material region.

The current–voltage characteristics were measured for MIS devices consisting of the Ti–TiO_2/PVDF composite and Si. The applied voltage range was −3 to 3 V. Figure 5a shows the leakage current and Figure 5b is the leakage current density of each sample. The asymmetric I–V relationship shown in Figure 5a is commonly observed in high-k dielectric materials. The asymmetry curve is correlated with the dielectric TiO_2 thickness [32]. The charge conduction under positive bias is believed to be dominated by the silicon/oxide interface, while that under negative bias is controlled by the metal/oxide interface [13]. As the annealing temperature of the samples increased, the leakage current density at Vg = 0 V decreased to 0.88 μA, 3.7 nA, and 2.0 pA for the 300, 450, and 600 °C samples in Figure 5b, respectively. This is consistent with the XRD and Raman spectroscopy results discussed above. The physical origin of this decrease in leakage current is the N_2 gas that flows during annealing. This is because, depending on the annealing temperature, N_2 is incorporated into the TiO_2 film, which helps to densify it and reduce the bulk and interfacial defect densities [33,34]. In particular, for the same thickness of the 600 °C sample, the results of this work show better leakage current characteristics than those reported in other studies [8,31]. These leakage current characteristics are related to the amount of residual PVDF, variation in the anatase phase, and N_2 incorporation as a function of heat treatment temperature.

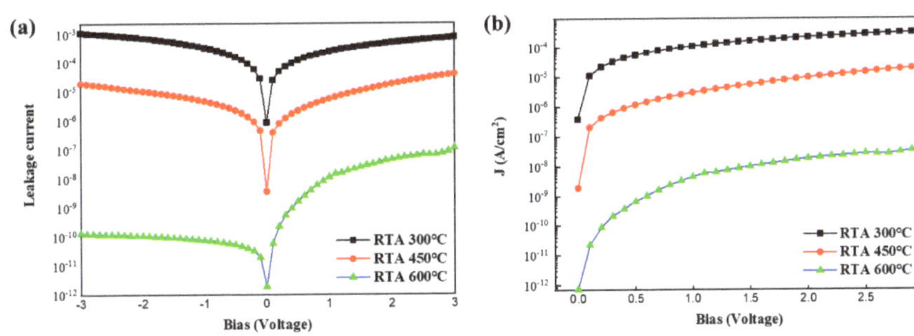

Figure 5. I–V characteristics of Ti–TiO_2/PVDF composite/Si MIS devices. (**a**) Current–voltage curve in voltage range of −3 to 3 V; (**b**) current density–voltage curve in voltage range of 0 to 3 V.

4. Conclusions

In this study, a dielectric thin film was fabricated on a Si wafer using a TiO_2/PVDF electrospinning solution with NMP as solvent. Generally, to fabricate MIS capacitors, each layer is deposited by CVD, ALD, and other techniques; however, in this study we used an electrospinning deposition method, which has never been reported before. Although electrospinning is commonly used to produce polymer nanofibers, in this study the dielectric layer was deposited by electrospraying. Compared to other deposition methods, electrospinning provides various advantages, such as a significantly shorter processing time, lower operating costs, and no requirement for vacuum equipment. In addition, the desired film thickness can be achieved by adjusting the processing time and, in the case of MIS capacitors, the optimal dielectric constant can be obtained by controlling the additive in the electrospinning solution.

Raman spectroscopy and XRD measurements showed that the fabricated thin films exhibited different leakage current characteristics depending on the annealing temperature, indicating a reduction in the residual PVDF amount and activation of the TiO_2 anatase phase. Among the fabricated thin films, the sample annealed at 600 °C showed the best leakage current (2.0 pA). This demonstrates the superior performance of TiO_2/PVDF composites as dielectric thin films. However, further research is needed to adjust the surface roughness and develop a process to form thin films with nanoscale thickness. As future work, we will also study the behavior of capacitance and cell potential of electrospun TiO_2 MIS.

To develop the next generation of capacitor devices, not only TiO_2 but also high-k materials such as HfO_2 can be added to the electrospinning solution, or the dielectric layer can be fabricated by mixing two or more materials with different k values, in order to match the dielectric constant to the application. The electrospinning-based coating is also one of the best methods for producing flexible films, as polymer solutions are utilized in the era of smaller scales and flexible electronic devices. Although electrospinning is still a process at the hundreds of nano- to multi-micro-scale, it has many advantages, and further research is required for its application in the semiconductor field.

Supplementary Materials: The following supporting information can be downloaded at https://www.mdpi.com/article/10.3390/mi15101231/s1. Figure S1: SEM image of sputtered Ti on the electrospun TiO_2 dielectric layer. Figure S2: The 5×5 μm^2 area AFM 3D topography (a) 300 °C, (b) 450 °C, and (c) 600 °C annealed samples. Figure S3: XRD data of an electrospun TiO_2 dielectric layer annealed at RTA temperatures of 300, 450, and 600 °C. (a) shows 2theta from 20° to 65°, (b) shows 2theta from 36° to 40°, (c) shows 2theta from 68.5° to 70.5°, and (d) shows 2theta from 25° to 26°. Figure S4: Raman data of an electrospun TiO_2 dielectric layer annealed at RTA temperatures of 300, 450, and 600 degrees. Table S1: Bond strength and surface roughness (R_a) as annealing temperature varies.

Author Contributions: Methodology, formal analysis, T.-M.C.; conceptualization, validation, J.-U.Y.; investigation, resources, writing—original draft, T.-M.C. and J.-U.Y.; writing—reviewing and editing, supervision, S.-G.P. All authors have read and agreed to the published version of the manuscript.

Funding: This work was supported by the Technology Innovation Program (or Industrial Strategic Technology Development Program Public–Private joint investment semiconductor R&D program (K-CHIPS) to foster high-quality human resources) ("RS-2023-00237003", High selectivity etching technology using cryoetch) funded by the Ministry of Trade, Industry and Energy (MOTIE, Republic of Korea) (1415187674). This work was also supported by the Technology Innovation Program (No. 20022472) funded by the Ministry of Trade, Industry and Energy (MOTIE, Republic of Korea). This research was supported by the Chung-Ang University research grant in 2023.

Data Availability Statement: The data presented in this study are available on request from the corresponding author.

Conflicts of Interest: The authors declare that they have no conflicts of interest.

References

1. Bourahla, N.; Hadri, B.; Bourahla, A. 3-D Simulation of Novel High Performance of Nano-Scale Dual Gate Fin-FET Inserting the High-K Dielectric TiO_2 at 5 Nm Technology. *Silicon* **2020**, *12*, 1301–1309. [CrossRef]
2. Li, Y.L.; Chang-Liao, K.S.; Li, C.C.; Huang, C.H.; Tsai, S.F.; Li, C.Y.; Hong, Z.Q.; Fang, H.K. Electrical and Reliability Characteristics of FinFETs with High-k Gate Stack and Plasma Treatments. *IEEE Trans. Electron Devices* **2021**, *68*, 4–9. [CrossRef]
3. Wei, C.Y.; Huang, M.Y.; Yan, S.C.; Wu, Y.C. High-Speed and Low-Power Ferroelectric HfO_2/ZrO_2 Superlattice FinFET Memory Device Using AlON Interfacial Layer. *IEEE Trans. Electron Devices* **2024**, *71*, 3977–3980. [CrossRef]
4. Yan, S.C.; Lan, G.M.; Sun, C.J.; Chen, Y.H.; Wu, C.H.; Peng, H.K.; Lin, Y.H.; Wu, Y.H.; Wu, Y.C. High Speed and Large Memory Window Ferroelectric $HfZrO_2$ FinFET for High-Density Nonvolatile Memory. *IEEE Electron Device Lett.* **2021**, *42*, 1307–1310. [CrossRef]
5. Jeon, W. Recent advances in the understanding of high-k dielectric materials deposited by atomic layer deposition for dynamic random-access memory capacitor applications. *J. Mater. Res.* **2020**, *35*, 775–794. [CrossRef]
6. Garzon-Roman, A.; Zuñiga-Islas, C.; Cuate-Gomez, D.H.; Heredia-Jimenez, A. TiO_2/porous silicon heterostructures formation by simple and low-cost methods for electronics applications. *Sens. Actuators A Phys.* **2023**, *349*, 114064. [CrossRef]
7. Shubham, K.; Khan, R. Annealing effect on fabrication and characterization of MIS structure using TiO_2-SiO_2 Thin Film as Insulator layer deposited by low temperature arc vapor deposition process. *J. Electron Devices* **2013**, *17*, 1439–1446.
8. Albertin, K.F.; Valle, M.; Pereyra, I. Study of MOS capacitors with TiO_2 and SiO_2/TiO_2 gate dielectric. *J. Integr. Circuits Syst.* **2007**, *2*, 89–93. [CrossRef]
9. Dittrich, T.; Weidmann, J.; Koch, F.; Lauermann, I.; Uhlendorf, I. Temperature- and oxygen partial pressure-dependent electrical conductivity in nanoporous rutile and anatase. *Appl. Phys. Lett.* **1999**, *75*, 3980. [CrossRef]
10. Kim, S.K.; Choi, G.-J.; Lee, S.Y.; Seo, M.; Lee, S.W.; Han, J.H.; Ahn, H.-S.; Han, S.; Hwang, C.S. Al-doped TiO_2 films with ultralow leakage currents for next generation DRAM capacitors. *Adv. Mater.* **2008**, *20*, 1429–1436. [CrossRef]
11. Yu, H.; Schaekers, M.; Schram, T.; Demuynck, S.; Horiguchi, N.; Barla, K.; Collaert, N.; Thean, A.V.; De Meyer, K. Thermal Stability Concern of Metal-Insulator-Semiconductor Contact: A Case Study of Ti/TiO_2/n-Si Contact. *IEEE Trans. Electron Devices* **2016**, *63*, 2671–2676. [CrossRef]
12. Masuda, Y.; Jinbo, Y.; Koumoto, K. Room Temperature CVD of TiO_2 Thin Films and Their Electronic Properties. *Sci. Adv. Mater.* **2009**, *1*, 138–143. [CrossRef]

13. Yang, W.; Marino, J.; Monson, A.; Wolden, C.A. An investigation of annealing on the dielectric performance of TiO_2 thin films, Semicond. *Sci. Technol.* **2006**, *21*, 1573. [CrossRef]
14. Kim, B.; Choi, Y.; Jeon, H.; Lee, D.; Byun, Y.; Jung, C. Leakage current characteristics of atomic layer deposited Al-doped TiO_2 thin film for dielectric in DRAM capacitor. *ECS J. Solid. State Sci. Technol.* **2021**, *10*, 083006. [CrossRef]
15. Wei, D.; Hossain, T.; Garces, N.; Nepal, N.; Meyer, H.; Kirkham, M.J.; Eddy, C.; Edgar, J.H. Influence of atomic layer deposition temperatures on TiO_2/n-Si MOS capacitor. *ECS J. Solid. State Sci. Technol.* **2013**, *2*, N110. [CrossRef]
16. Seo, M.; Rha, S.H.; Kim, S.K.; Han, J.H.; Lee, W.; Han, S.; Hwang, C.S. The mechanism for the suppression of leakage current in high dielectric TiO_2 thin films by adopting ultra-thin HfO_2 films for memory application. *J. Appl. Phys.* **2011**, *110*, 024105–24111. [CrossRef]
17. Kim, J.Y.; Jung, H.S.; No, J.H.; Kim, J.-R.; Hong, K.S. Influence of anatase-rutile phase transformation on dielectric properties of sol-gel derived TiO_2 thin films. *J. Electroceram.* **2006**, *16*, 447–451. [CrossRef]
18. Yoo, J.U.; Kim, D.H.; Choi, T.M.; Jung, E.S.; Lee, H.R.; Lee, C.Y.; Pyo, S.G. Advancements in Flexible Nanogenerators: Polyvinylidene Fluoride-Based Nanofiber Utilizing Electrospinning. *Molecules* **2024**, *29*, 3576. [CrossRef]
19. Huang, A.; Liu, F.; Cui, Z.; Wang, H.; Song, X.; Geng, L.; Peng, X. Novel PTFE/CNT composite nanofiber membranes with enhanced mechanical, crystalline, conductive, and dielectric properties fabricated by emulsion electrospinning and sintering. *Compos. Sci. Technol.* **2021**, *214*, 108980. [CrossRef]
20. Wang, H.; Zhang, Y.; Niu, H.; Wu, L.; He, X.; Xu, T.; Wang, N.; Yao, Y. An electrospinning–electrospraying technique for connecting electrospun fibers to enhance the thermal conductivity of boron nitride/polymer composite films. *Compos. B Eng.* **2022**, *230*, 109505. [CrossRef]
21. Mokhtari, F.; Latifi, M.; Shamshirsaz, M. Electrospinning/electrospray of polyvinylidene fluoride (PVDF): Piezoelectric nanofibers. *J. Text. I.* **2016**, *107*, 1037. [CrossRef]
22. Costa, L.M.M.; Bretas, R.E.S.; Gregorio, R. Effect of solution concentration on the electrospray/electrospinning transition and on the crystalline phase of PVDF. *Mater. Sci. Appl.* **2010**, *1*, 246–251. [CrossRef]
23. Hu, J.; Zhang, S.; Tang, B. Rational design of nanomaterials for high energy density dielectric capacitors via electrospinning. *Energy Storage Mater.* **2021**, *37*, 530–555. [CrossRef]
24. Wan, B.; Li, H.; Xiao, Y.; Yue, S.; Liu, Y.; Zhang, Q. Enhanced dielectric and energy storage properties of $BaTiO_3$ nanofiber/polyimide composites by controlling surface defects of $BaTiO_3$ nanofibers. *Appl. Surf. Sci.* **2020**, *501*, 144243. [CrossRef]
25. Wang, R.; He, J.; Yan, C.; Jing, R.; Zhao, Y.; Yang, J.; Shi, M.; Yan, X. A Long-Range Planar Polymer with Efficient π-Electron Delocalization for Superior Proton Storage. *Adv. Mater.* **2023**, *36*, 2402681. [CrossRef]
26. Jin, J.; Wang, R.; Yu, K.; Tao, Y.; Zhang, P.; Ke, L.; Yang, J.; Shi, M. Imine-based conjugated polymer enables efficient removal of ammonium ion via capacitive deionization. *Sep. Purif. Technol.* **2025**, *353*, 128290. [CrossRef]
27. Yoo, J.-U.; Kim, D.-H.; Jung, E.-S.; Choi, T.-M.; Lee, H.-R.; Pyo, S.-G. Enhanced Hybrid Nanogenerator Based on PVDF-HFP and PAN/BTO Coaxially Structured Electrospun Nanofiber. *Micromachines* **2024**, *15*, 1171. [CrossRef]
28. Chen, C.; Shen, L.; Liu, G.; Cui, Y.; Yan, S. Improved Energy Storage Performance of Composite Films Based on Linear/Ferroelectric Polarization Characteristics. *Polymers* **2024**, *16*, 1058. [CrossRef]
29. Thamaphat, K.; Limsuwan, P.; Ngotawornchai, B. Phase characterization of TiO_2 powder by XRD and TEM. *Agric. Nat. Resour.* **2008**, *42*, 357–361.
30. Kang, B.; Choi, K.-K.; An, J.; Baek, R.-H. Demonstration of TiO_2 Based Ultra High-k (k = 30) Metal-Insulator–Semiconductor Capacitor and Its Electrical Properties. *J. Nanosci. Nanotechnol.* **2021**, *21*, 4394–4399. [CrossRef]
31. Li, W.; Ni, C.; Lin, H.; Huang, C.; Shah, S.I. Size dependence of thermal stability of TiO_2 nanoparticles. *J. Appl. Phys.* **2004**, *96*, 6663–6668. [CrossRef]
32. Lin, J.-Y.J.; Roy, A.M.; Sun, Y.; Saraswat, K.C. In Metal-insulator-semiconductor contacts on Ge: Physics and applications. In Proceedings of the 2012 International Silicon-Germanium Technology and Device Meeting (ISTDM), Berkeley, CA, USA, 4–6 June 2012; pp. 1–2.
33. Chakraborty, S.; Bera, M.K.; Bhattacharya, S.; Maiti, C.K. Current conduction mechanism in TiO_2 gate dielectrics. *Microelectron. Eng.* **2005**, *81*, 188–193. [CrossRef]
34. Sun, S.C.; Chen, T.F. A novel approach for leakage current reduction of LPCVD Ta_2O_5 and TiO_2 films by rapid thermal N_2O annealing. In Proceedings of the 1994 IEEE International Electron Devices Meeting, Electron Devices Meeting, Yokohama, Japan, 23–26 August 1994; pp. 333–336. [CrossRef]

Disclaimer/Publisher's Note: The statements, opinions and data contained in all publications are solely those of the individual author(s) and contributor(s) and not of MDPI and/or the editor(s). MDPI and/or the editor(s) disclaim responsibility for any injury to people or property resulting from any ideas, methods, instructions or products referred to in the content.

micromachines

Article

Study of Weight Quantization Associations over a Weight Range for Application in Memristor Devices

Yerim Kim [1], Hee Yeon Noh [1], Gyogwon Koo [2], Hyunki Lee [2], Sanghan Lee [3], Rock-Hyun Choi [2], Shinbuhm Lee [4], Myoung-Jae Lee [1,3] and Hyeon-Jun Lee [1,3,*]

1 Division of Nanotechnology, Daegu Gyeongbuk Institute of Science and Technology (DGIST), Daegu 42988, Republic of Korea
2 Division of Intelligent Robot, Daegu Gyeongbuk Institute of Science and Technology (DGIST), Daegu 42988, Republic of Korea
3 Institute of Next-Generation Semiconductor Convergence Technology (INST), Daegu Gyeongbuk Institute of Science and Technology (DGIST), Daegu 42988, Republic of Korea
4 Department of Physics and Chemistry, Daegu Gyeongbuk Institute of Science and Technology (DGIST), Daegu 42988, Republic of Korea
* Correspondence: dear.hjlee@dgist.ac.kr

Abstract: The development of hardware-based cognitive computing systems critically hinges upon the integration of memristor devices capable of versatile weight expression across a spectrum of resistance levels while preserving consistent electrical properties. This investigation aims to explore the practical implementation of a digit recognition system utilizing memristor devices with minimized weighting levels. Through the process of weight quantization for digits represented by 25 or 49 input signals, the study endeavors to ascertain the feasibility of digit recognition via neural network computation. The integration of memristor devices into the system architecture is poised to streamline the representation of the resistors required for weight expression, thereby facilitating the realization of neural-network-based cognitive systems. To minimize the information corruption in the system caused by weight quantization, we introduce the concept of "weight range" in this work. The weight range is the range between the maximum and minimum values of the weights in the neural network. We found that this has a direct impact on weight quantization, which reduces the number of digits represented by a weight below a certain level. This was found to help maintain the information integrity of the entire system despite the reduction in weight levels. Moreover, to validate the efficacy of the proposed methodology, quantized weights are systematically applied to an array of double-layer neural networks. This validation process involves the construction of cross-point array circuits with dimensions of 25 × 10 and 10 × 10, followed by a meticulous examination of the resultant changes in the recognition rate of randomly generated numbers through device simulations. Such endeavors contribute to advancing the understanding and practical implementation of hardware-based cognitive computing systems leveraging memristor devices and weight quantization techniques.

Keywords: memristor; neural network; weight quantization; recognition

check for updates

Citation: Kim, Y.; Noh, H.Y.; Koo, G.; Lee, H.; Lee, S.; Choi, R.-H.; Lee, S.; Lee, M.-J.; Lee, H.-J. Study of Weight Quantization Associations over a Weight Range for Application in Memristor Devices. *Micromachines* **2024**, *15*, 1258. https://doi.org/10.3390/mi15101258

Academic Editors: Chengyuan Dong and Kuibo Yin

Received: 25 July 2024
Revised: 30 September 2024
Accepted: 14 October 2024
Published: 15 October 2024

Copyright: © 2024 by the authors. Licensee MDPI, Basel, Switzerland. This article is an open access article distributed under the terms and conditions of the Creative Commons Attribution (CC BY) license (https://creativecommons.org/licenses/by/4.0/).

1. Introduction

Artificial intelligence finds application across diverse domains such as life sciences, communication, and transportation, rooted in the concept of mimicking brain cells, or neurons. Neural networks, the cornerstone of artificial intelligence, operate on a signaling system where stimulation passes from one cell to another [1]. The current artificial intelligence systems necessitate computerized learning and processing to address a myriad of challenges, encompassing tasks like pattern recognition and real-time processing. However, the traditional sequential processing, following the von Neumann architecture, often results in bottlenecks [2] due to the varying processing speeds of the devices. A novel approach proposes hardware [3] implementation of artificial intelligence to overcome these limitations. Recent studies comparing the efficiency of a software-based neural network [4] and a Convolutional Neural Network with hardware implementations have shown hardware to be more efficient [5]. This hardware-based artificial intelligence, mimicking human brain function, is termed neuromorphic technology [6].

The previous neuromorphic research efforts have explored innovative approaches, such as a study that devised a circuit utilizing memristors [7,8] to transfer the "weights" of a neural network [9]. Furthermore, other studies have tackled tasks like learning and recognizing a 3 × 3 alphabet by employing existing Convolutional Neural Networks (CNNs) represented as an assembly of eight-layer memristors, both in simulations and real-world implementations [10,11]. However, a notable limitation of such studies lies in the challenge of configuring the network settings in hardware-based experiments, particularly in implementing crucial factors like "weight". In one instance, despite conducting the experiments with a simple dataset, difficulties arose in implementing the weights, leading to their conversion into binary or being represented in bit units using multiple memristors. Alternative research endeavors have explored expressing weight as variable resistance [12], although this method faces constraints in replicating the artificial intelligence performance obtained via software. Consequently, a novel technique known as "Quantization" has emerged, offering a simplified approach to handling weights [13].

Quantization involves representing the input values from a continuous range as discrete values. Another study utilized this approach to construct a model by modifying the activation function of an existing network [14,15]. For application in physical hardware, the weight levels need to be minimized, prompting the previous studies to explore binary methods utilizing memristors and circuits with variable resistors [11,12,16]. In addition, there are studies that quantized the states of LRS and HRS in order to utilize the device, and cases in which the efficiency was improved by fitting the correlation between pulse and conductance [17,18]. However, these methods suffer from reduced accuracy due to limitations in expressing numerous weights. Even in software-based implementations, binary models exhibit suboptimal accuracy [19]. To address these issues, in this work, we introduce a technique to properly quantize weights, and we propose a method to optimize them to improve accuracy. The quantization of weights is optimized by introducing the concept of "weight range", and it is successful to reduce the values of more than three-hundred-fifty different weights to four weights. In quantization, variables are expressed in a limited number, and, at this time, how small the number of variables is that can be expressed is important. In order to control this, the concept of weight range was used, and it was found that there is a correlation between the distribution of the corresponding value and the expression value during quantization. Using this, a method to express the variables of a neural network with a small number of numbers was proposed. In addition, we discuss the system implementation of a neural network using a resistance-based memristor device.

2. Methods

In neural network research aimed at developing artificial intelligence, complex networks of connections between neurons are built through recurrent and regression operations on inputs and outputs. The objective is to generate an output based on a given input by leveraging numerous neurons to execute operations and transmit signals to subsequent neurons. This process involves a technique known as "backpropagation" [1], wherein the network is trained by propagating the error of the computed value to achieve the optimal learning outcome. When performing weight updates via backpropagation, the large number of weight values that need to be integrated in individual memristor elements in an array structure makes it difficult to implement cognitive computing systems. With memristors in current stage [8,20–22], it is not easy to create more than 100 weight levels, and the stability of each level is unstable. Therefore, this research focuses on minimizing the number of weight levels that the trained model must represent through quantization. In this particular study, initial weights were assigned random values, and the final weights were determined through backpropagation. All computations were carried out using Python (version 3.11), with signal propagation achieved solely through addition, subtraction, and multiplication operations without utilizing any pre-defined functions. The Python version used to construct the neural network was 3.11.3, and the network was built using only the Numpy and Random packages. All operations were performed on arrays in Numpy, and sigmoid [1,23] was used as the activation function. The mean squared error method was used as the loss function in the backpropagation method [24]. This equation sets the difference between the expected value and the actual correct answer as the error and learns by correcting the square of the difference. Learning was conducted with 100 times iteration per epoch, and training was completed for 5000 epochs in total.

To validate the functionality of the computed weights on an actual device, Cadence's electrical design automation (EDA), a semiconductor process simulator, was employed to verify the electrical signals of the physical device. To verify the actual device's behavioral characteristics, simulations were performed using Cadence Virtuso 6.1.7. For the process design kit (PDK), we used the NSPL CMOS 0.5 µm process in the INST (Institute of Next-generation Semiconductor convergence Technology, DGIST) nano-Fab. To implement the simulation conditions, we performed mappings in the form of (weight, resistance), (input, voltage), and (output, current). For the input voltage, we added 0.65 V to account for the threshold voltage (*Vth*) of the diode used in the program. The output current was also used with the addition of a reference resistor element to keep it at a constant level. Conventional memristor devices can be used as devices for cognitive computation, and in this study, we performed PDK evaluation to implement a system by implementing resistors and diodes using Si. A quantized resistor device was fabricated and its electrical characteristics (Ketheley 2636b) and local thermal analysis (Nanoscopesystems TRM250) were measured, but only the results of PDK are shown in this study.

3. Results and Discussion

3.1. Weight Quantizaiton

To implement a system capable of cognitive computation utilizing the memristors researched to date, memristors capable of representing at least 100 or more weights are needed, and these memristors must be stable to drive and reproducible enough to distinguish the resistance between each weight level. However, the current memristors are not directly applicable to the implementation of cognitive computing systems due to the limited number of weight levels, retention problems, and endurance problems. Various problems have been raised with these memristor devices, such as the migration of oxygen atoms [25], changes in the electrical properties through the diffusion of hydrogen [26,27], the destruction of the device in localized areas by pulse signals [28], and problems with the uniformity of the memristor filaments [29]. If the number of weight levels that the model needs to represent can be minimized through quantization techniques, cognitive computing systems using memristor devices can be made easier.

The experimental dataset utilized both 5 × 5 and 7 × 7 numeric inputs, with a binary representation using 0 and 1. The training consisted of randomly entering ten numbers from zero to nine into a 5 × 5 or 7 × 7 grid cell. Each number was represented by a 1 in the corresponding grid cell (Figure 1a,b). In the case of the 5 × 5 input, marking '2' would be depicted as ([1,1,1,1,1], [0,0,0,0,1], [1,1,1,1,1], [1,0,0,0,0,0], [1,1,1,1,1]) (Figure 1d). Subsequently, the data were flattened and fed into a single input array. The 5 × 5 input initially entered the input layer as a (1 × 25) array and underwent sigmoid computation with a weight1 of size (25 × 10) and an activation function, resulting in a hidden layer of size (1 × 10). The data then proceeded through another layer with weight2 of size (10 × 10) before undergoing sigmoid computation once more. The output layer provided a probability value ranging from zero to one for each digit, selecting the highest probability and converting it into a (1 × 10) output where one element was represented by 1. For the 7 × 7 numeric input, the input data took the form of (1 × 49), with the subsequent weights determined as (49 × 10). The succeeding hidden layers followed the same format as the

5×5 input, with sizes of (1×10) for the first layer and (10×10) for the second layer. Among these parameters, the values in the input layer remained fixed throughout the backpropagation process, with only weight1 and weight2 being quantized subsequently. During the quantization, weight1 and weight2 were not treated individually but instead combined. This combined set of weight1 (W1) and weight2 (W2) is denoted by W.

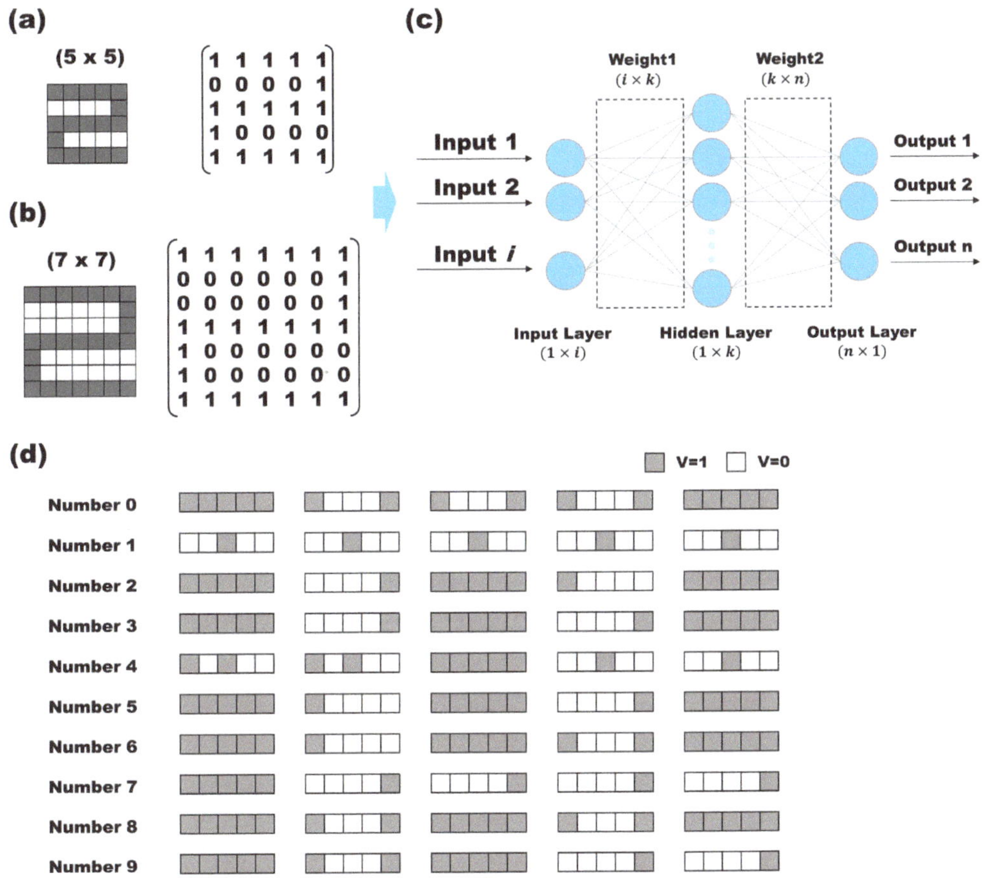

Figure 1. In a neural network experiment, two inputs are introduced: (**a**) a (5×5) array representing numbers from 0 to 9, and (**b**) a (7×7) array. The signal represented in the array is injected into the system in the form of a matrix. (**c**) The system used is a double neural network layer with one hidden layer. (**d**) The 10 numbers used as input were converted into 25 data.

Figure 2a illustrates the flowchart detailing the optimization and quantization process for each weight within a 5×5 input dataset. In the unconstrained case, achieving a 100% recognition rate for randomly input values necessitated approximately 500,000 times the backpropagation for the values of W1 and W2. We initially trained the model with 100% accuracy because the 5×5 and 7×7 data are relatively simple datasets consisting of 0s and 1s, so the accuracy can easily reach 100% if the training volume is sufficient. After sufficient training, we moved on to the next step, quantization. In this process, we only used W, which stands for W1 and W2 among the model parameters, to perform the quantization. The subsequent quantization process aimed to represent a continuum of values divided into specific steps. The key components for implementing this include the "weight range"

(Equation (1)) and a "level (q)" for quantization. q is a number more than 0 and has a range below p, which is the maximum value.

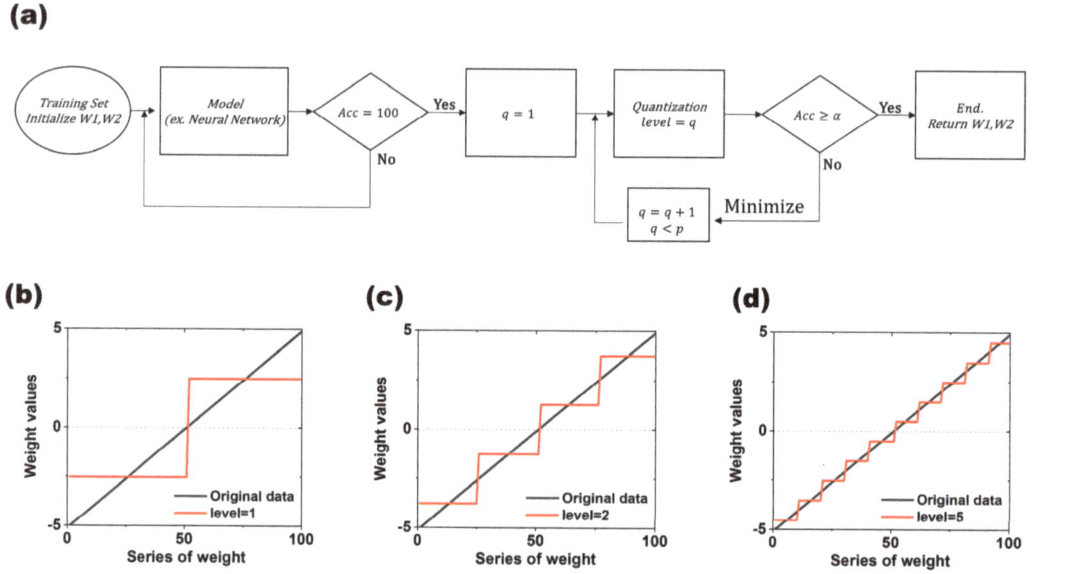

Figure 2. Weight quantization and process flow: (**a**) W1 and W2 fed through the training set are trained via backpropagation and move to the next step when they reach 100% recognition. They sequentially increase the quantization level (q) to find weights that satisfy α. (**b**) Conceptual diagram of weight quantization at quantization level 1, (**c**) at level 2, and (**d**) at quantization level 5.

Figure 2b–d are the result of the quantization dependent on level (q) (q = 1, 2, and 4). The black line is the original value, and the red line is the value after the quantization. As the level (q) increases, the number of steps increases, which makes the difference between the values before and after the quantization smaller. The level (q) serves as the target value for converting weights from their original analog form to the discrete values aimed for during quantization. The weight range represents the difference between the minimum and maximum values of W, expressed as the sum of W1 and W2 (Equation (2)). The interval is defined as the weight range divided by the desired level (k), indicating the range of values represented by each single level (Equation (3)). The value quantizing the actual range value at $level_{(i,\ middle)}$ is then determined by multiplying the interval by a value of $i + 0.5$ and $\min(W_{total})$ (Equation (4)).

$$Weight\ range = \max(W_{total}) - \min(W_{total}) \tag{1}$$

$$Weight\ rangeA = \max(|W_{total}|) - \min(|W_{total}|) \tag{2}$$

$$interval = (Weight\ rangeA\ )/(level\ (q)\ ) \tag{3}$$

$$level_{i,\ middle} = \min(W_{total}) \times interval \times (i + 0.5)\ , \quad i = 1,\ 2,\ \cdots,\ p\ (0 < q < p)\ , \tag{4}$$

where, i is the number of levels. Equations (1), (3), and (4) are used when the list of values being quantized consists of only positive numbers greater than zero. In this case, "Weight rangeA" in Equation (3) and "Weight range" in Equation (1) have the same value because all the numbers are positive. Conversely, if the list of values being quantized contains both positive and negative numbers, "Weight rangeA" is used to set the interval.

The initial weight values used in each experiment were used as seed values to consider 60 cases, and these values were randomly generated. The process of generating and learning a new initial weight was repeated 100 times for each individual seed value. Thus, 6000 models were created for each dataset, with each model subsequently quantized to assess the results. The quantization was performed by starting with $q = 1$ and incrementing the value of q sequentially until an optimal value of q was obtained that satisfied α (α is the target recognition rate accuracy). To proceed with the quantization while ensuring accuracy, if the recognition rate accuracy value is greater than or equal to the value of α, move to the next step; otherwise, q is added to 1 and the quantization is performed again. In this experiment, α was set to 100%, and, in general, the smaller α is, the lower the level value that can be reached. The quantization process runs under the condition that $q < p$, with the maximum value set to p, and the quantization process terminates when q exceeds p, as shown in the flowchart in Figure 2. Although all the points of $level_{(i,\ middle)}$ are established by p, there are cases where $level_{(i,\ middle)}$ is not generated in a certain interval because the number of W does not exist between those intervals. In this study, the non-generated cases were not considered when determining the final level.

After training the neural network based on the 5 × 5 and 7 × 7 numeric input datasets, quantization was performed based on the calculated weights. Of the parameters generated from the 5 × 5 numeric dataset, only W1 and W2 were used for the quantization. There are 350 numbers in W as an array, including 250 W1 and 100 W2. An array of W generated from a 7 × 7 number dataset has 590 numbers. The weight quantization procedure outlined in Figure 2 was implemented for the 5 × 5 dataset. Following the division of the 350 weights generated by the 5 × 5 dataset into four levels, the recognition rate remained unchanged (Figure 3). This signifies a substantial reduction of 98.9% in the weight levels compared to their original state. Similarly, for the 590 variables in the 7 × 7 dataset, a reduction of 99.3% was observed, resulting in a reduction of four levels. Figure 3 illustrates the distribution of the 350 weights derived from a single 5 × 5 dataset. The x-axis is arranged in ascending order of weights, while the y-axis represents the corresponding weight values at each point. These weights vary from 0 to 12.5, with a prevalence of lower values. The dashed black line represents the unconstrained weight distribution, exhibiting 350 distinct values. By segmenting the 350 weight values into four levels, the quantized weights, depicted in red in Figure 3, were derived. The first weight level condenses the use of 296 distinct weight values into a single value, while the second, third, and fourth levels amalgamate 30 or fewer weight values into a singular value. It is noteworthy that, while many of the values are close to zero, the levels are created not only for these values but also for those with lesser numerical significance. Specifically, the distribution containing 296 items exhibits a high distribution of values, whereas the distribution with only five items demonstrates a notably lower distribution. It is important to note that not all 6000 models employed in this experiment underwent quantization with four weights. Only in some datasets was it computationally observed to maintain a 100% recognition rate even after reducing the weights to four levels. Based on the experimental results, the stable reduction in the weight levels varied depending on the magnitude of the maximum and minimum values of the weights, which aligns with the definition of "Weight range" explained earlier.

3.2. Weight Range

When designing and implementing actual neural network systems in hardware, the challenge is to implement them while maintaining the accuracy of the various weights. Quantization, which is used to compensate for this drawback, has the advantage of reducing the complexity and implementation difficulty when implemented in hardware (using a memristor device). When designing a neuromorphic model, the neural network model must be finite, and, in the process, all the parameters must be represented by elements, which increases the processing difficulty of the elements, increases the complexity, and increases the cost of implementation. An algorithm used to address this is quantization, which is commonly used in the field of reducing model complexity and increasing performance.

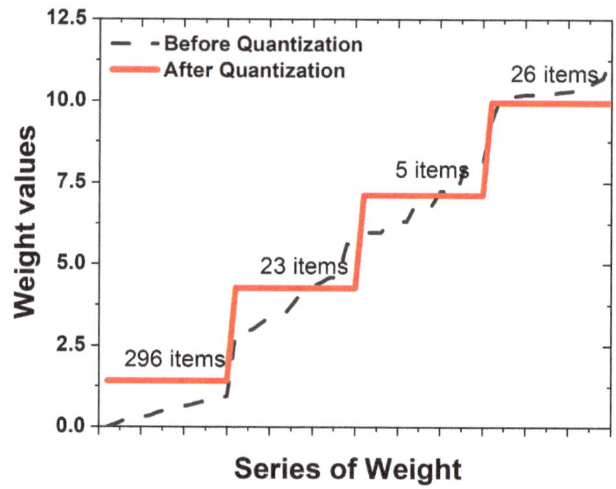

Figure 3. Distribution of weight values and quantization levels: (dashed line) distribution of weight values before the quantization process; (red line) quantization of 350 different weight values into 4 values.

The existing quantization technique offers the advantage of converting the single-precision floating point (FP32) to a half-precision floating point (FP16) or normalizing the layer [14]. This technique reduces the number precision by decreasing the calculation bits while maintaining the range, aiming to cut the memory use and computational complexity. Split into FP32 and FP16, FP32 employs 32 bits, while FP16 uses 16 bits. FP16 offers reduced memory and increased throughput in software but minimal advantages in hardware, where FP32 and FP16 provide no practical benefits. To address this, our proposed quantization technique emphasizes actual hardware device implementation, simplifying the model's weight into a straightforward list of values. Notably, our study focuses on minimizing the unique weights by setting a low level and conducting quantization. Therefore, in order to reduce the level, the correlation was investigated using the concept of weight range mentioned above. Figure 4a presents the outcomes of an experiment aimed at determining the optimal levels for various weight ranges. The initial weights were seeded from one to sixty, and 100 iterations were conducted to generate 6000 distinct models for 5 × 5 and 7 × 7 inputs, respectively. The correlation between the weights employed in each model and the resulting number of levels in the final weights was examined. The number of levels in the final weights was defined as the point at which the recognition rate reached 100%, denoted as q when α is 100% in Figure 2a. Notably, the minimum level achievable for the quantized levels (q) appears to be directly proportional to the weight range. For both the 5 × 5 and 7 × 7 input systems, models with α satisfying 100% are found at $q = 4$, and it has been observed that weight ranges have significantly lower values of q around 21. An interesting aspect is the phenomenon of having the same weight range but different values of q level. This suggests that there are other factors that determine the level of q along with the weight range. For models with a weight range greater than 60, the value of q is determined between 10 and 40. The points A (8, 32) and B (25, 62), shown in Figure 4a, were plotted in Figure 4b for a representative model with weight ranges of 32 and 62. As already mentioned in Figure 3, the weights of the 6000 models used in this experiment are mostly clustered around zero, and this trend becomes more pronounced as the weight range increases.

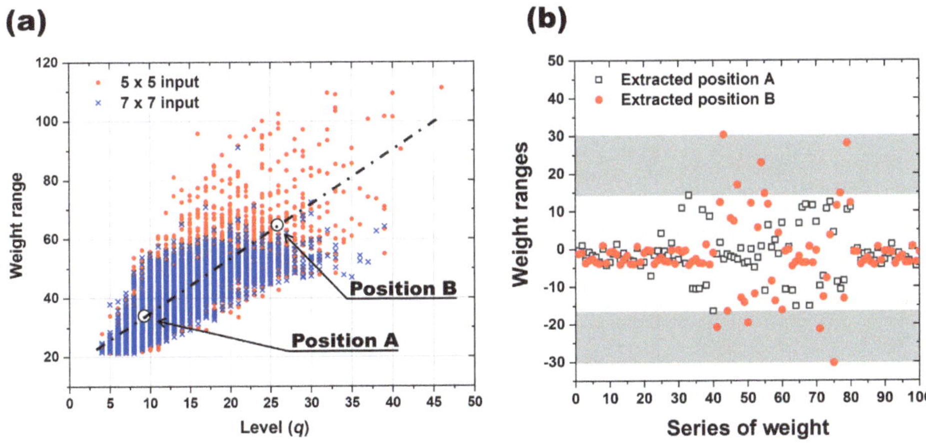

Figure 4. Distribution and influence of weight range: (**a**) influence of the number of quantized levels (k) on the distribution of weight range; (**b**) distribution of two different weight ranges, A and B.

In this comprehensive analysis encompassing 6000 models, a significant correlation between "weight range" and "q", the quantizable level, was established. For the same input signal, the weight range can be an important factor in determining q. In addition, for the system satisfying quantization level q = 4, the change in the recognition rate as a function of the weight range over the number of epochs was calculated as shown in Figure 5. The relationship between quantization level (q) and weight range was studied by comparing the post-quantization recognition rate accuracy for 10 randomly selected models in different training states using five seed values. Figure 5a illustrates a graph depicting the distribution of the training weight ranges, where 100 models were generated for each "Seed". The figure shows a sequential increase in both the mean value and distribution of the weight ranges from Seed A to Seed E. These seeds are integral for refining the initial weights and facilitating the subsequent training. The outcomes of computing the recognition rate accuracy based on the initial weights categorized by their respective weight ranges for Seeds A, C, and E are presented in Figure 5b–d, respectively. The accuracy of the recognition rate was evaluated as the number of computations increased for each of the three seeds in a system satisfying quantization level q = 4. Notably, for 'Seed A', characterized by the smallest weight range, the recognition rate accuracy demonstrated a consistent upward trend with each epoch, exceeding 90% accuracy after 400 epochs, as depicted in Figure 5b. 'Seed C' showed the same trend of gradual improvement in recognition rate accuracy as 'Seed A' as the number of epochs increased, with a recognition rate accuracy of about 70% after 600 epochs. After that, there was no further improvement in the recognition rate with increasing epochs, as shown in Figure 5c. In contrast to Seeds A and C, Seed E demonstrated a notable absence of improvement in the recognition rate with the advancement of training epochs. Initiating at the initial accuracy level of 20%, it maintained this accuracy level throughout the duration, even at epoch 1000, as illustrated in Figure 5d. This evaluation was performed by artificially implementing seeds with a very large range of weight ranges, which shows that the weight range quantization level, q, of the weight can have a direct impact on the recognition rate.

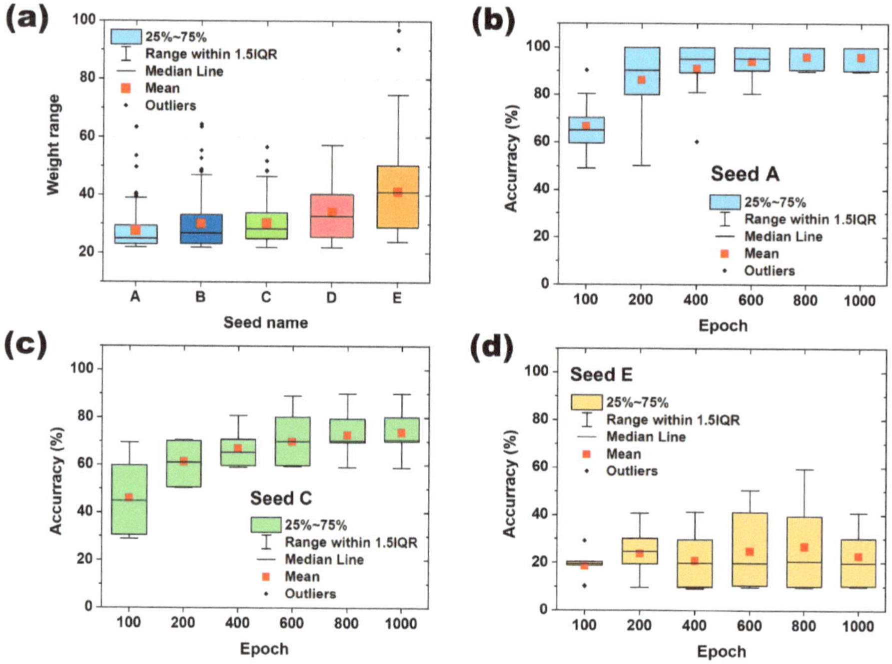

Figure 5. Classification of "seeds" according to weight range and the recognition rate accuracy for each seed: (**a**) configure seeds with different weight ranges in 5 zones, where the weight range is gradually increased from A to E. (**b**–**d**) show the change in recognition rate as the neural network is trained on Seed A, Seed C, and Seed E, respectively.

3.3. Circuit Implementation

So far, it has been shown through neural network calculations that a 100%-digit recognition rate can be obtained by using the relationship between weight range and quantization level q. The device simulation for an actual device chipset was completed via PDK. The PDK simulation was performed by selecting the weight that converges to the lowest q level of 4 in the computational calculation. The circuit is shown in Figure 6a, where the input is represented by voltage and the output is represented by current. Compared to Figure 1, I is mapped to i, K to k, and N to n. Between layers 1 and 2, a reference resistor is used to convert the current to voltage and to compensate for the current. In Figure 6, the reference resistors used for each line are all the same. To check the influence of the threshold voltage of the diode added to prevent sneak current, the difference between the recognition rate and the output current as a function of the magnitude of the input voltage was compared through circuit simulation. Figure 6b shows the accuracy of the recognition rate for the final output of the PDK simulation and the difference between the current computed by the neural network and the PDK computation. The accuracy shows the digit recognition rate for 10 different inputs (0–9), with a 100% recognition rate for all the input values. This experiment shows that even a system consisting of only four weights, which is reduced to four by quantizing the weights when building a chip that runs on a real device, can still provide reliable digit recognition. The rate of the current gap shown in Figure 6b represents the difference between the output current from the PDK simulator for device fabrication and the computational calculation for the neural network. The graph shows that the input number "9" has the rate of current gap 27% (red line in Figure 6b, using a 1 V input voltage). It can be seen that input numbers 0, 5, and 6, including 9, have more than 10% gaps regarding the output value compared to the computational neural network calculation. These gaps are the difference between the calculated output current value and the current value in the PDK simulation for driving the actual device, so they do not contribute much to the decrease in recognition rate. This trend was similar when increasing the data representing the input numbers (increasing the input information in 5 × 5 inputs). This shows that the output current value maintains a certain margin, ensuring accuracy even in different environments (gaps generated by random sources; in this case, gaps generated by device fabrication).

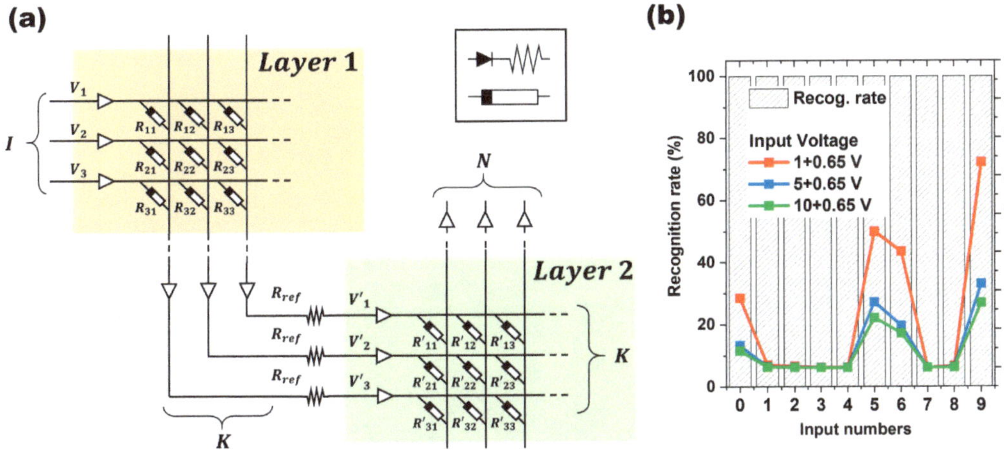

Figure 6. (**a**) A schematic diagram of the circuit to drive the actual device using PDK simulation; (**b**) the rate of the final current gap value from the neural network computational calculation and the number recognition rate of the device obtained from the PDK simulator.

The difference in the current values between the neural network calculation and the PDK simulation is due to the diodes used as selectors. The diode used in this study has a threshold voltage of 0.65 V, so, if the input voltage is similar to this, the threshold voltage of the diode will be affected. As it goes through the circuits of W1 and W2, the value of the output current for each matrix slightly deviates from the calculated value, and the difference is clearly revealed in the final output current. The blue and green lines in Figure 6b show "the current gap" when the input voltage is increased to 5 V and 10 V, respectively. It can be seen that the ratio of the current gap gradually decreases as the input voltage increases. In a computer calculation, the output current is (V-Vth)/R, but, in a PDK simulation, the voltage actually changes across the resistor after it passes the diode's threshold voltage and undergoes feedback in the circuit.

Quantization of weights in neural network research refers to the process of reducing the precision of the weights in a neural network. In neural networks, weights are typically represented as floating-point numbers, which require a certain amount of memory and computational resources to store and process. In general, quantization aims to reduce the memory and computational requirements of a neural network by representing the weights with fewer bits. The application of quantization to simple inputs (a 5 × 5 or 7 × 7 matrix) in this study is aimed at reducing the computational requirements of the neural network's memory and reducing the number of weights that need to be implemented by memristors as much as possible in the application of hardware-based systems, thus facilitating the implementation of device-based neural network systems (Figure 7). Quantizing weights is essential for deploying neural networks on resource-constrained devices such as mobile phones, IoT devices, and embedded systems. However, quantization can lead to model inaccuracy, which requires careful optimization and tuning to mitigate. To quantize the weights, we introduced the concept of weight range, which provides the possibility to adjust the number of levels that can be quantized. We evaluated the cognitive operation for relatively simple digit recognition using 5 × 5 (25 inputs) and 7 × 7 (49 inputs) inputs and found a 100% recognition rate, but further evaluation of the applicability of this system to large input data is needed. Also, quantization can lead to model inaccuracies, which require careful optimization and tuning to mitigate.

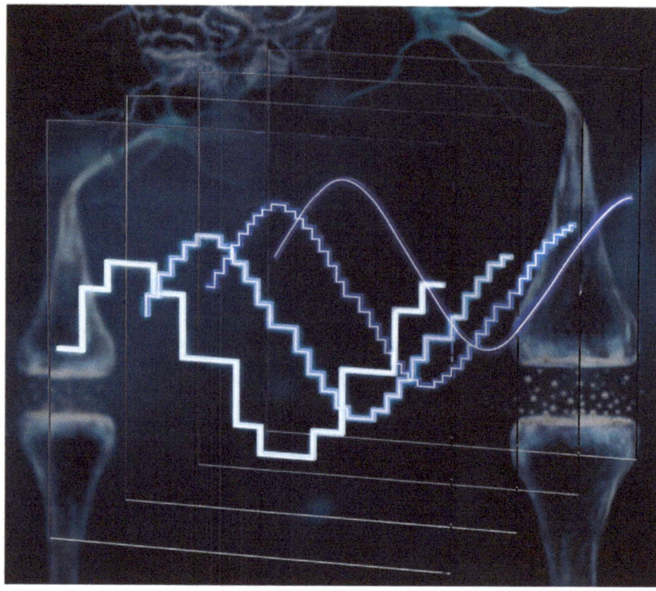

Figure 7. In object recognition via cognitive operations, a large amount of weights containing traditional decimal forms are simplified by quantization. Quantized weights not only reduce the amount of memory in neural network computations but can also reduce the number of weight levels that a device must represent in neural network computations implemented with memristor devices.

4. Conclusions

We investigated weight quantization within the context of neural network computation and device simulation. Weight quantization was applied to compare the recognition rates in a dual-layer neural network structure using simple numeric input signals consisting of 5 × 5 and 7 × 7 matrices. In particular, we introduced the concept of weight ranges to optimize the quantization efficiency, demonstrating its correlation with weight quantization and the ensuing recognition rates. It was shown that, by simplifying the weights while maintaining the accuracy above a certain value set during quantization, the number of numbers that need to be represented by weights can be reduced by more than 95% while maintaining a certain level of recognition rate. The device simulations were performed by applying the quantized weights to an array of two neural network layers, and it was shown that the number recognition rate can be stably secured even in actual device operation.

Overall, the efforts to implement neural network computation with memristor devices are still ongoing. However, the implementation of memristor-based neural network computation systems utilizing the devices currently being researched is not easy due to the instability of the devices, high stability, and the requirement for functionality under many conditions. Advancing the quantization technology presented in this study is expected to enable the implementation of artificial intelligence systems with high energy efficiency. In particular, a prototype memristor-based cognitive computing system for simple systems will be realized through further research.

Author Contributions: Y.K. fabricated/measured and calculated the device and drafted the manuscript. H.Y.N., G.K. and H.L. contributed to the calculational concept and discussion of the results. S.L. (Sanghan Lee) contributed to the device simulation, and R.-H.C., S.L. (Shinbuhm Lee) and M.-J.L. contributed to the overall understanding of the device physics and provided comments on the theoretical mechanisms involved. H.-J.L. contributed to the designed experimental concept and discussed the device physics and concepts. All authors have read and agreed to the published version of the manuscript.

Funding: This research was supported in part by the DGIST R&D Program of the Ministry of Science (24-CoE-NT-01, 24-ET-02), and in part by the National Research Foundation of Korea (NRF) grant funded by the Korea Government (MSIT) (No. 2023R1A2C1004131), and in part by the Commercialization Promotion Agency for R&D Outcomes (COMPA) funded by the Ministry of Science and ICT (MSIT). (RS-2024-00417600).

Data Availability Statement: The data presented in this study are available upon request from the corresponding author.

Acknowledgments: The authors would like to provide thanks for device design by Cadence from Institute of Next-generation Semiconductor convergence Technology (INST), Korea. Special thanks to June-Seo Kim and Chan-Kang Lee at the DGIST, and Ji-Young Woo at Kyeon-Buk National University for their valuable technical assistance.

Conflicts of Interest: The authors declare no competing interests.

References

1. Schmidhuber, J. Deep learning in neural networks: An overview. *Neural Netw.* **2015**, *61*, 85–117. [CrossRef] [PubMed]
2. Nei, M.; Maruyama, T.; Chakraborty, R. The Bottleneck Effect and Genetic Variability in Populations. *Evolution* **1975**, *29*, 1–10. [CrossRef] [PubMed]
3. Su, F.; Liu, C.; Stratigopoulos, H.G. Testability and Dependability of AI Hardware: Survey, Trends, Challenges, and Perspectives. *IEEE Des. Test* **2023**, *40*, 8–58. [CrossRef]
4. Gao, J.; Heng, F.; Yuan, Y.; Liu, Y. A novel machine learning method for multiaxial fatigue life prediction: Improved adaptive neuro-fuzzy inference system. *Int. J. Fatigue* **2024**, *178*, 108007. [CrossRef]
5. Sze, V.; Chen, Y.H.; Yang, T.J.; Emer, J.S. Efficient Processing of Deep Neural Networks: A Tutorial and Survey. *Proc. IEEE* **2017**, *105*, 2295–2329. [CrossRef]
6. Mead, C. Neuromorphic electronic systems. *Proc. IEEE* **1990**, *78*, 1629–1636. [CrossRef]
7. Lee, M.-J.; Lee, C.B.; Lee, D.; Lee, S.R.; Chang, M.; Hur, J.H.; Kim, Y.-B.; Kim, C.-J.; Seo, D.H.; Seo, S.; et al. A fast, high-endurance and scalable non-volatile memory device made from asymmetric Ta2O5−x/TaO2−x bilayer structures. *Nat. Mater.* **2011**, *10*, 625–630. [CrossRef]
8. Haripriya, G.R.; Noh, H.Y.; Lee, C.-K.; Kim, J.-S.; Lee, M.-J.; Lee, H.-J. Interface roughness effects and relaxation dynamics of an amorphous semiconductor oxide-based analog resistance switching memory. *Nanoscale* **2023**, *15*, 14476–14487. [CrossRef]
9. Yao, P.; Wu, H.; Gao, B.; Tang, J.; Zhang, Q.; Zhang, W.; Yang, J.J.; Qian, H. Fully hardware-implemented memristor convolutional neural network. *Nature* **2020**, *577*, 641–646. [CrossRef]
10. Lin, P.; Li, C.; Wang, Z.; Li, Y.; Jiang, H.; Song, W.; Rao, M.; Zhuo, Y.; Upadhyay, N.K.; Barnell, M.; et al. Three-dimensional memristor circuits as complex neural networks. *Nat. Electron.* **2020**, *3*, 225–232. [CrossRef]
11. Prezioso, M.; Merrikh-Bayat, F.; Hoskins, B.D.; Adam, G.C.; Likharev, K.K.; Strukov, D.B. Training and operation of an integrated neuromorphic network based on metal-oxide memristors. *Nature* **2015**, *521*, 61–64. [CrossRef] [PubMed]
12. Kim, S.; Gokmen, T.; Lee, H.M.; Haensch, W.E. Analog CMOS-based resistive processing unit for deep neural network training. In Proceedings of the 2017 IEEE 60th International Midwest Symposium on Circuits and Systems (MWSCAS), Boston, MA, USA, 6–9 August 2017; pp. 422–425. [CrossRef]
13. Gray, R.M.; Neuhoff, D.L. Quantization. *IEEE Trans. Inf. Theory* **1998**, *44*, 2325–2383. [CrossRef]
14. Gholami, A.; Kim, S.; Zhen, D.; Yao, Z.; Mahoney, M.; Keutzer, K. A Survey of Quantization Methods for Efficient Neural Network Inference. In *Low-Power Computer Vision*, 1st ed.; Chapman and Hall: London, UK; CRC: Boca Raton, FL, USA, 2022; pp. 291–326.
15. Yang, J.; Shen, X.; Tian, X.; Li, H.; Deng, B.; Huang, J.; Hua, X. Quantization Networks. In Proceedings of the 2019 IEEE/CVF Conference on Computer Vision and Pattern Recognition (CVPR), Long Beach, CA, USA, 15–20 June 2019; pp. 7308–7316.
16. Xu, J.; Huan, Y.; Yang, K.; Zhan, Y.; Zou, Z.; Zheng, L.R. Optimized Near-Zero Quantization Method for Flexible Memristor Based Neural Network. *IEEE Access* **2018**, *6*, 29320–29331. [CrossRef]
17. Zhang, Y.; Wu, Z.; Liu, S.; Guo, Z.; Chen, Q.; Gao, P.; Wang, P.; Liu, G. A Quantized Convolutional Neural Network Implemented With Memristor for Image Denoising and Recognition. *Front. Neurosci.* **2021**, *15*, 717222. [CrossRef]
18. Mehonic, A.; Joksas, D.; Ng, W.H.; Buckwell, M.; Kenyon, A.J. Simulation of Inference Accuracy Using Realistic RRAM Devices. *Front. Neurosci.* **2019**, *13*, 593. [CrossRef]
19. Tang, W.; Hua, G.; Wang, L. How to Train a Compact Binary Neural Network with High Accuracy? In Proceedings of the 31st AAAI Conference on Artificial Intelligence, San Francisco, CA, USA, 4–9 February 2017. [CrossRef]

20. Li, H.-L.; Su, J.; Xu, M.-H.; Dong, S.-J.; Bian, J.; Shan, P.-S.; Wang, R.-W.; Liu, Y.; Wang, X.-L.; Fan, S.-Q.; et al. Resistance switching stability of STO memristor under Au ion implantation. *Appl. Phys. Lett.* **2024**, *124*, 013505. [CrossRef]
21. Lee, M.-J.; Park, G.-S.; Seo, D.H.; Kwon, S.M.; Lee, H.-J.; Kim, J.-S.; Jung, M.; You, C.-Y.; Lee, H.; Kim, H.-G.; et al. Reliable Multivalued Conductance States in TaOx Memristors through Oxygen Plasma-Assisted Electrode Deposition with in Situ-Biased Conductance State Transmission Electron Microscopy Analysis. *ACS Appl. Mater. Interfaces* **2018**, *10*, 29757–29765. [CrossRef]
22. Naqi, M.; Yu, Y.; Cho, Y.; Kang, S.; Khine, M.T.; Lee, M.; Kim, S. Integration of IGZO-based memristor and Pt-based temperature sensor for enhanced artificial nociceptor system. *Mater. Today Nano* **2024**, *27*, 100491. [CrossRef]
23. Han, J.; Moraga, C. The influence of the sigmoid function parameters on the speed of backpropagation learning. In *From Natural to Artificial Neural Computation*; Mira, J., Sandoval, F., Eds.; Springer: Berlin/Heidelberg, Germany, 1995; pp. 195–201.
24. Rumelhart, D.E.; Hinton, G.E.; Williams, R.J. Learning representations by back-propagating errors. *Nature* **1986**, *323*, 533–536. [CrossRef]
25. Su, R.; Xiao, R.; Shen, C.; Song, D.; Chen, J.; Zhou, B.; Cheng, W.; Li, Y.; Wang, X.; Miao, X. Oxygen ion migration induced polarity switchable SrFeOx memristor for high-precision handwriting recognition. *Appl. Surf. Sci.* **2023**, *617*, 156620. [CrossRef]
26. Noh, H.Y.; Kim, J.; Kim, J.-S.; Lee, M.-J.; Lee, H.-J. Role of Hydrogen in Active Layer of Oxide-Semiconductor-Based Thin Film Transistors. *Crystals* **2019**, *9*, 75. [CrossRef]
27. Noh, H.Y.; Lee, W.-G.; Haripriya, G.R.; Cha, J.-H.; Kim, J.-S.; Yun, W.S.; Lee, M.-J.; Lee, H.-J. Hydrogen diffusion and its electrical properties variation as a function of the IGZO stacking structure. *Sci. Rep.* **2022**, *12*, 19816. [CrossRef] [PubMed]
28. Lee, H.; Abe, K. A Study on the Effect of Pulse Rising and Falling Time on Amorphous Oxide Semiconductor Transistors in Driver Circuits. *IEEE Electron Device Lett.* **2020**, *41*, 896–899. [CrossRef]
29. Wang, H.; Yan, X.; Wang, S.; Lu, N. High-Stability Memristive Devices Based on Pd Conductive Filaments and Its Applications in Neuromorphic Computing. *ACS Appl. Mater. Interfaces* **2021**, *13*, 17844–17851. [CrossRef]

Disclaimer/Publisher's Note: The statements, opinions and data contained in all publications are solely those of the individual author(s) and contributor(s) and not of MDPI and/or the editor(s). MDPI and/or the editor(s) disclaim responsibility for any injury to people or property resulting from any ideas, methods, instructions or products referred to in the content.

micromachines

Article

Application of Solution-Processed High-Entropy Metal Oxide Dielectric Layers with High Dielectric Constant and Wide Bandgap in Thin-Film Transistors

Jun Liu [1], Xin Xiong [2], Han Li [3], Xiangchen Huang [1], Yajun Wang [1], Yifa Sheng [1,*], Zhihao Liang [2], Rihui Yao [2,4], Honglong Ning [2,5] and Xiaoqin Wei [3,*]

1 School of Electrical Engineering, University of South China, Hengyang 421001, China; 2021000103@usc.edu.cn (J.L.); 2020000113@usc.edu.cn (Y.W.)
2 Guangdong Basic Research Center of Excellence for Energy & Information Polymer Materials, State Key Laboratory of Luminescent Materials and Devices, School of Materials Sciences and Engineering, South China University of Technology, Guangzhou 510640, China
3 Southwest Institute of Technology and Engineering, Chongqing 400039, China
4 Key Lab of Guangdong Province for High Property and Functional Polymer Materials, South China University of Technology, Guangzhou 510640, China
5 The International School of Microelectronics, Dongguan University of Technology, Dongguan 523808, China
* Correspondence: sheng@usc.edu.cn (Y.S.); weixiaoqin810913@163.com (X.W.)

Abstract: High-k metal oxides are gradually replacing the traditional SiO_2 dielectric layer in the new generation of electronic devices. In this paper, we report the production of five-element high entropy metal oxides (HEMOs) dielectric films by solution method and analyzed the role of each metal oxide in the system by characterizing the film properties. On this basis, we found optimal combination of $(AlGaTiYZr)O_x$ with the best dielectric properties, exhibiting a low leakage current of 1.2×10^{-8} A/cm^2 @1 MV/cm and a high dielectric constant, while the film's visible transmittance is more than 90%. Based on the results of factor analysis, we increased the dielectric constant up to 52.74 by increasing the proportion of TiO_2 in the HEMOs and maintained a large optical bandgap (>5 eV). We prepared thin film transistors (TFTs) based on an $(AlGaTiYZr)O_x$ dielectric layer and an $InGaZnO_x$ (IGZO) active layer, and the devices exhibit a mobility of 18.2 cm^2/Vs, a threshold voltage (V_{th}) of −0.203 V, and an subthreshold swing (SS) of 0.288 V/dec, along with a minimal hysteresis, which suggests a good prospect of applying HEMOs to TFTs. It can be seen that the HEMOs dielectric films prepared based on the solution method can combine the advantages of various high-k dielectrics to obtain better film properties. Moreover, HEMOs dielectric films have the advantages of simple processing, low-temperature preparation, and low cost, which are expected to be widely used as dielectric layers in new flexible, transparent, and high-performance electronic devices in the future.

Keywords: high-k; high entropy metal oxide; solution method; TFTs

check for updates

Citation: Liu, J.; Xiong, X.; Li, H.; Huang, X.; Wang, Y.; Sheng, Y.; Liang, Z.; Yao, R.; Ning, H.; Wei, X. Application of Solution-Processed High-Entropy Metal Oxide Dielectric Layers with High Dielectric Constant and Wide Bandgap in Thin-Film Transistors. *Micromachines* **2024**, *15*, 1465. https://doi.org/10.3390/mi15121465

Academic Editors: Ai-Qun Liu and Chengyuan Dong

Received: 31 October 2024
Revised: 27 November 2024
Accepted: 29 November 2024
Published: 30 November 2024

Copyright: © 2024 by the authors. Licensee MDPI, Basel, Switzerland. This article is an open access article distributed under the terms and conditions of the Creative Commons Attribution (CC BY) license (https://creativecommons.org/licenses/by/4.0/).

1. Introduction

Accompanied by the rapid development of science and technology, the development of electronic devices with high integration ability, high response speed, and low power consumption has become a general trend. In this process, the dielectric layer, one of the important components of various electronic devices, has been put forward with low leakage current (<10^{-8} A), high dielectric constant (k > 10), and high transmittance (>85%) requirements [1]. With the miniaturization and high integration ability of electronic devices, the traditional SiO_2 dielectric layer (with its low dielectric constant) requires reduced thickness, which leads to increased tunneling effect and leakage current. Therefore, it is gradually being replaced by high-k materials (k > 10) [2]. The main factors that researchers focus on when selecting high-K materials include dielectric constant, band gap width, crystallinity, surface morphology, and defect states [3,4]. Currently, a variety of promising high-k dielectric materials have received attention, such as Al_2O_3 [5], Ga_2O_3 [6], HfO_2 [7,8], TiO_2 [9], Y_2O_3 [10–12], ZrO_2 [13–15], and so on. Among them, HfO_2, TiO_2, and ZrO_2 have the advantage of high dielectric constant (k > 20), but there are problems such as high leakage current and small forbidden bandgap (<5 eV) [16]. Al_2O_3, as a common metal oxide dielectric material, is favored because of the advantages of stable performance, good insulating properties, and high crystallization temperature (>800 °C) [17], but the disadvantage of low dielectric constant (k < 10) makes it unable to meet the requirements of high integration of electronic devices. It can be seen that a single metal oxide faces challenges in balancing multiple desirable properties for effective practical application.

By mixing or doping, various high-k materials can be combined to make up for the shortcomings of a single material system, taking into account the performance advantages of a variety of materials, to meet the requirements of electronic devices for high-performance dielectric layers with low leakage current, high dielectric constant, and high transmittance [16,18–23]. There have been many studies on the combination of binary or ternary metal oxides, which can be targeted to optimize one of the properties such as dielectric constant, leakage current, or forbidden bandgap. High-entropy metal oxides (HEMOs) contain five or more elements, promising a flexible combination of high-performance metal oxide dielectric materials [24]. High entropy materials are loosely defined as solid solution materials containing more than five dominant elements with equal or nearly equal atomic percentages. The concept of high entropy opens up a new path for the development of advanced materials with unique properties that cannot be achieved by traditional materials approaches based on only one dominant element [25]. Meanwhile, each atom in HEMOs is surrounded by neighboring atoms of different sizes with a highly asymmetric disordered structure, which increases the probability of carrier-atom collisions, effectively reduces the carrier leakage pathway, leading to better dielectric properties [26,27]. Therefore, in this paper, six high-performance dielectric materials with their respective advantages, Al_2O_3 for the low leakage current, Ga_2O_3 for the wide forbidden bandgap, HfO_2, TiO_2, ZrO_2 for the high dielectric constant and Y_2O_3 as carrier inhibitor, were selected to be combined, and six sets of five-element high-entropy combinations were designed for the study, namely, $(GaHfTiYZr)O_x$, $(AlHfTiYZr)O_x$, $(AlGaTiYZr)O_x$, $(AlGaHfYZr)O_x$, $(AlGaHfTiZr)O_x$, $(AlGaHfTiY)O_x$. The solution method is selected for the preparation of five-element high-entropy thin films, which not only has the advantages of a simple process and low cost but also is very convenient for realizing the mixing of multiple elements as well as the proportion regulation [28].

2. Materials and Methods

2.1. Precursor Solution Preparation and Characterization

2-Methoxyethanol (2-MOE) was chosen as the precursor solvent, and the precursor solutes were selected as aluminum nitrate hydrate ($Al(NO_3)_3{\cdot}9H_2O$), gallium nitrate hydrate ($Ga(NO_3)_3{\cdot}xH_2O$), hafnium chloride ($HfCl_4$), tetrabutyl titanate ($C_{16}H_{36}O_4Ti$), yttrium nitrate hexahydrate ($Y(NO_3)_3{\cdot}6H_2O$), and zirconium nitrate ($Zr(NO_3)_4{\cdot}5H_2O$), five of which were selected for combination at a time, to configure a high-entropy precursor solution with a concentration of 1.0 M. The molar ratios of the five metal salts were 1:1:1:1:1. We renamed the six-element combinations for clarity, the nomenclature is shown in Table 1. Further optimization we performed by varying the concentration of Ti ions in the AGTYZ fraction. The solvent and solute of the solution were the same as mentioned earlier, but the concentration of the $C_{16}H_{36}O_4Ti$ concentrations were adjusted. The total solute concentration in the solution was 1.0 M and the concentration of Ti was increased from 0 M to 0.28 M while the other four solutes were decreased from 0.25 M to 0.18 M. 1% acetic acid was added to the solution to inhibit the hydrolysis of the metal ions. Acetic acid was chosen to avoid the effect of the added acid on the film composition because it's easy to be removed by heating. After stirring the precursor solution for 12 h, we filtered it with a 13 mm/0.45 μm filter and left it to age for 24 h. The surface tension of the solutions was measured using an optical surface tension meter (Attension T200, Biolin Scientific, Gothenburg, Sweden) and the viscosity was studied using a rheometer (HAAKE MARS 40, Thermo Fisher Scientific, Waltham, MA USA). TG-DSC tests were conducted using a thermogravimetric analyzer (DZ-TGA101, Nanjing Dazhan, Nanjing, China) and a differential scanning calorimeter (DZ-DSC300C, Nanjing Dazhan, Nanjing, China).

Table 1. HEMOs sample nomenclature.

Sample Names	**(GaHfTiYZr)O_x**	**(AlHfTiYZr)O_x**	**(AlGaTiYZr)O_x**	**(AlGaHfYZr)O_x**	**(AlGaHfTiZr)O_x**	**(AlGaHfTiY)O_x**
Abbreviation	GHTYZ	AHTYZ	AGTYZ	AGHYZ	AGHTZ	AGHTY

2.2. *Film Fabrication and Characterization*

The films were deposited on quartz glass substrates by spin-coating (1000 r/min for 6 s then 5000 r/min for 30 s) with 50 μL of precursor. Next, the films were pre-annealed in the air at 150 °C for 10 min and then annealed at 200 °C, 300 °C, and 400 °C for 1.5 h, respectively. The surface morphology of the films was observed by a laser scanning confocal microscopy (OLS5000-CB, Olympus, Tokyo, Japan) and an atomic force microscope (BY3000, Nano Instruments, Beijing, China). The phase of the films was analyzed using an X-ray diffractometer (EMpyrean X, PANalytical, Almelo, The Netherlands). The optical properties of the films were investigated by a UV-Vis spectrophotometer (UV-2600 Shimadzu, Fukuoka ken, Japan). The thickness of the films was measured by a stylus profiler (Dektak 150, Bruker, Billerica, MA, USA).

2.3. *Metal-Insulator-Metal Device Fabrication and Characterization*

The electrical properties of HEMOs films were characterized using metal-insulator-metal (MIM) devices with an ITO/HEMOs/Al structure, where HEMOs films were deposited on ITO substrates by spin-coating, and 100-nm-thick Al top electrodes were deposited using thermal evaporation. The device structure is visualized in Figure 1. The current-voltage (I-V), capacitance-frequency (C-F), and capacitance-voltage (C-V) characteristics of the MIM devices were characterized by a semiconductor parameter analyzer (FS-Pro, Primarius Technologies, Shanghai, China).

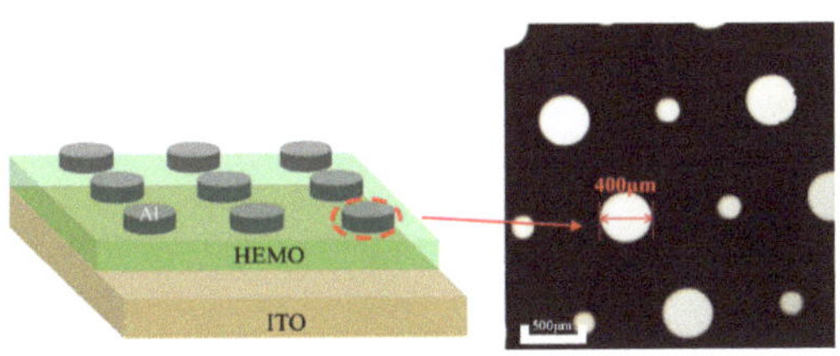

Figure 1. Schematic diagram of MIM device structure.

2.4. *Thin-Film Transistor Applications and Characterization*

HEMOs films were first spin-coated onto T-shaped Al substrates. Subsequently, 20-nm-thick IGZO active layers were deposited via RF pulse sputtering, followed by thermal annealing at 250 °C for 1 h, then 100-nm-thick Al source-drain electrodes deposited through thermal evaporation. Output curves and transfer curves of the TFTs were measured with a semiconductor parameter analyzer (FS Pro, Primarius Technologies, Shanghai, China).

2.5. *The Critical Parameter of Thin-Film Transistor*

Vth corresponds to the value of the gate voltage when a conductive channel is formed at the active/insulating layer interface. Vth can be obtained by intersecting the Vg axis with the epitaxial part of the linear portion of the $I_d^{1/2}$. Vth can be obtained by intersecting the Vg axis with the epitaxial part of the linear part of the Id-to-Vg curve in the transfer characteristic.

Mobility (μ) is related to the efficiency of carrier transport in semiconductors and directly affects the maximum current and operating frequency of the device. Mobility can be calculated in different ways. This article calculates the saturation mobility and the formula is shown in (1):

$$\mu = \frac{2L\left(\frac{d\sqrt{I_d}}{dVg}\right)^2}{C_i W} \quad (1)$$

The switching ratio (I_{on}/I_{off}) is defined as the ratio of the open-state current (I_{on}) to the closed-state current (I_{off}), and the ratio of the maximum current to the minimum current in the transfer characteristic curve is taken in the actual calculation. The inverse value at the maximum slope on the transfer characteristic curve is called SS, as shown in equation (2).

$$SS = \frac{\partial Vg}{\partial \log(I_d)} \quad (2)$$

3. Results and Discussion

3.1. Solution Properties

3.1.1. Rheological Properties

The rheological properties of the solutions significantly impact the quality of spin-coated film [29]. The results of surface tension (ST) characterization are shown in Figure 2a–g. The ST were slightly decreased compared with the theoretical value of ethylene glycol methyl ether (31.3 mN/m), suggesting that the metal salts reduced the solution's surface tension. However, varying combinations of elements had minimal effects on the surface tension. As shown in Figure 2g, the ST of all six solutions were lower than 30 mM/m, the solution viscosities were lower than 5 mPa·s, and the rheological properties of the precursor solutions were in the range suitable for the spin-coating method [30].

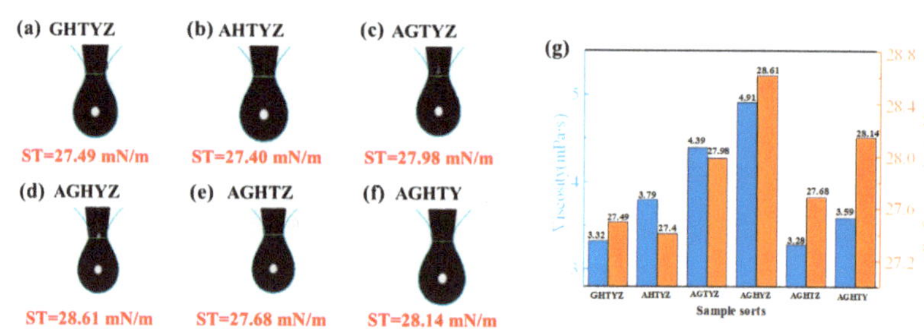

Figure 2. (**a–f**) Results of solution ST tests; (**g**) Surface tension and viscosity of precursor solutions.

3.1.2. Chemical Properties

The solutions were subjected to TG-DSC tests, and the results are shown in Figure 3a–f. The TG curves of the solutions showed similar trends. They had an obvious mass loss in the temperature interval from room temperature to 125 °C, which was due to the rapid evaporation of the solvent, ethylene glycol methyl ether. In addition, it could be observed that the TG curves exhibited inflection points between 100 °C and 125 °C, accompanied by a decreased rate of mass loss. This phenomenon can be attributed to the significant increase in solution concentration after substantial solvent evaporation, strengthening the binding force between the remaining solvent molecules and the solute molecules, thus requiring more energy for molecules to escape from the solution system [31,32]. Above 125 °C, the mass loss rate decreased, which may result from the continued decomposition of organic residuals and acid groups [33]. The presence of heat absorption peaks around 50 °C in some curves may be due to desorption of adsorbed moisture [34,35]. The endothermic peaks around 150 °C observed in some curves may be associated with the hydrolysis of metal cations forming hydrogen bonds and the formation of reticulated structures through metal ion coordination. Based on the TG-DSC analysis, the optimal annealing temperature should exceed 150 °C. Changes in the components of the solution during heating were detected by TGDSC tests. Lower temperatures may result in incomplete organic solvent removal, hindering the formation of dense and uniform films, while excessive temperatures could promote crystallization, increasing device leakage current. Therefore, the annealing temperatures were first set to 200 °C, 300 °C, and 400 °C to further optimize the temperature.

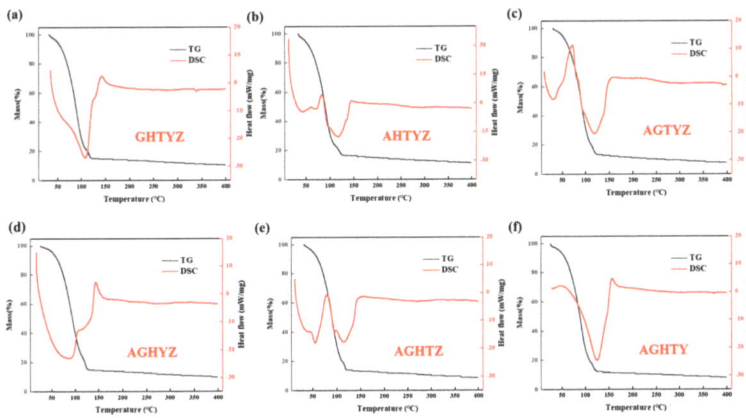

Figure 3. TG-DSC characterization results of solutions: (**a**) GHTYZ solution; (**b**) AHTYZ solution; (**c**) AGTYZ solution; (**d**) AGHYZ solution; (**e**) AGHTZ solution; (**f**) AGHTY solution.

3.2. *Films Properties*

3.2.1. Physical Properties

The film thickness characterized by the stylus profiler and the roughness measured by the AFM are shown in Figures 4a and 4b respectively. The film thickness and surface roughness exhibit similar trends with the change of components. While temperature is higher than 200 °C, except for the AGHTZ and AGHTY, the film thickness ranged from 70 to 80 nm, and the Root Mean Square of the roughness (Sq) of the films is lower than 2 nm. At an annealing temperature of 200 °C, the films generally show increased thickness, especially GHTYZ and AGHTY. As the annealing temperature increases to 300 °C and 400 °C, the film thickness stabilizes, indicating that temperature ceases to be the dominant factor affecting film thickness above 300 °C. At this time, the AGHTY film still shows abnormally high film thickness, combined with the film roughness test results shown in Figure 4b, the AGHTY film also shows abnormally high surface roughness (Sq > 20 nm).

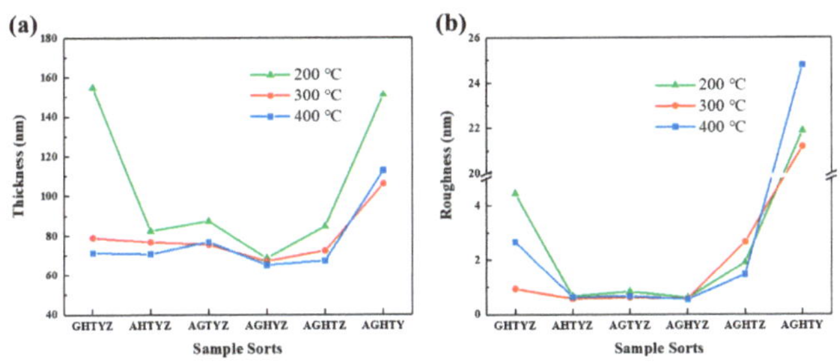

Figure 4. (**a**) Thickness of HEMOs films; (**b**) roughness of HEMOs films.

The films annealed at different temperatures were subjected to XRD tests as shown in Figure 5a–c. Only the diffraction peaks of the glass substrate can be measured around $2\theta \approx 21°$, which indicates that all the high-entropy oxide films are mainly amorphous even when the annealing temperature is up to 400 °C. According to some previous studies, the crystallization temperature of ZrO_2 and HfO_2 is low (about 350 °C), while Al_2O_3 and Ga_2O_3 have high crystallization temperatures. Therefore, the additions of Al_2O_3 and Ga_2O_3 can effectively increase the overall crystallization temperature of the films. Amorphous film is favorable for large-area preparation and can inhibit the formation of current channels, which helps reduce device leakage current and device power consumption [36]. Based on the comprehensive analysis of both the solutionand film properties, different annealing temperatures were investigated. At 200 °, organic residuals and acid ions can not be completely removed from the films, degrading the film density and uniformity. When the annealing temperature was increased from 300 °C to 400 °C, no significant improvement in film morphology was observed. Considering both the cost-effectiveness of the preparation process and the requirement of low-temperature deposition for flexible electronics, an annealing temperature of 300 °C was selected for further experiments.

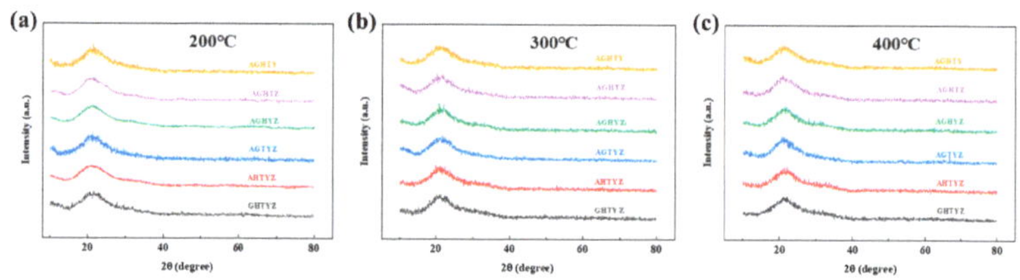

Figure 5. XRD test results of HEMOs films with different annealing temperatures: (**a**) 200 °C; (**b**) 300 °C; (**c**) 400 °C.

Here the physical properties of the film have been analyzed, such as film thickness, roughness and crystallinity. From the test results, except for AGHTYZ and annealing temperatures above 200 °C, the component films obtain a good surface property and are in an amorphous state.

3.2.2. Optical Properties

The optical properties of the high entropy metal oxide films were characterized by a UV-Vis spectrophotometer. The transmittance of the films are shown in Figure 6a. All the films containing Zr show good visible light transmittance (over 89% in the 380–380 nm range), which shows their potential to be used in transparent electronic devices. As shown in Figure 6b, the optical band gaps were derived by linear fitting the $(ah\nu)^2$-$h\nu$ curves. The optical bandgap of the films is shown in Figure 6c. Since the high-entropy system contains a variety of materials with high forbidden bandgap, the HEMOs films exhibit large optical bandgaps (>5 eV), with the AGHYZ films having the largest optical bandgap (~5.63 eV). The bandgap of TiO_2 (~3.5 eV) is the smallest among six oxides, thus the optical bandgap of the HEMOs films containing TiO_2 is reduced.

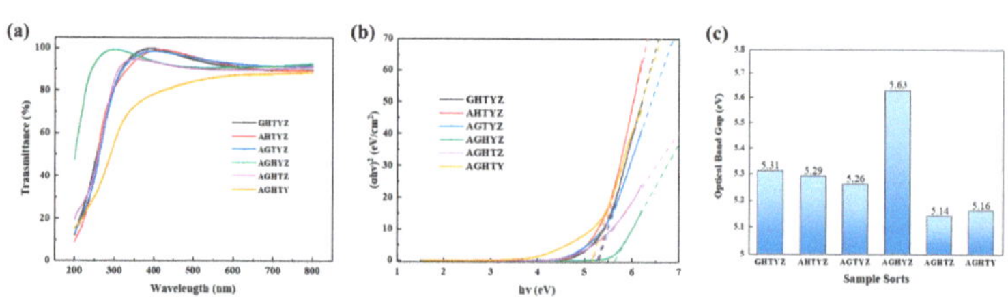

Figure 6. Results of film optical characterization: (**a**) transmittance; (**b**) $(ah\nu)^2$-$h\nu$ curves; (**c**) optical bandgap of HEMOs films.

The films were characterised by high transmittance and high bandwidth through transmittance testing and bandwidth fitting, which facilitates the obtaining of a dielectric film with a low leakage current.

3.2.3. Components Analysis

XPS test was conducted to acquire information on elemental content and chemical states. XPS analysis of the AGTYZ film revealed characteristic peaks for Al2p, Ga3d, Ti2p, Y3d, and Zr3d, along with the percentage of each element in the AGTYZ, as shown in Figure 7a,b. The observed valence states of these 5 metal elements correspond to the typical oxidation states in their common oxides. A similar situation occurs with AHTYZ films, as shown in Figure 7c,d. The distribution of the six metal oxides in the films has been obtained by XPS of these two compositions. It showed that each element constitutes approximately 20% of the film composition, consistent with the intended stoichiometry. This confirms that multi-elemental HEMOs dielectric layers can be successfully synthesized via solution processing, and the proportion of each content can be easily controlled.

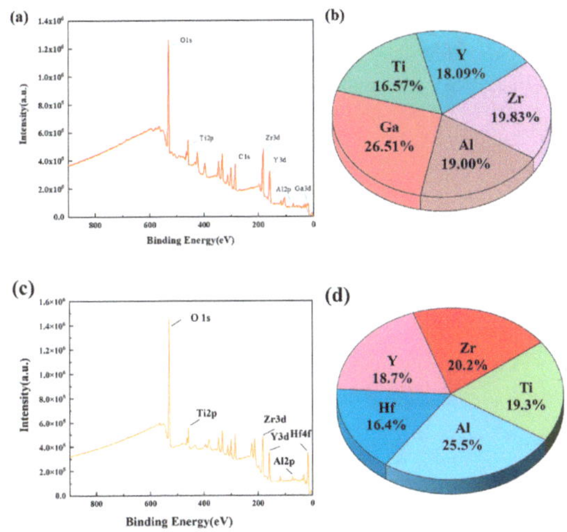

Figure 7. (**a**) XPS full spectrum of the AGTYZ film; (**b**) the percentage of each element in the AGTYZ film; (**c**) XPS full spectrum of the AHTYZ film; (**d**) the percentage of each element in the AHTYZ film.

The O1s XPS spectra were deconvoluted into three peaks with a Gaussian distribution, as shown in Figure 8a. The peaks centered at 529.7 eV, 531.1 eV, and 532.2 eV correspond to metal–oxygen bonds (M-O-M), oxygen vacancies (V_o), and hydroxyl groups (M-OH), respectively, formed through water adsorption and related processes [37]. The proportions of the three oxygen chemical states in HEMO films are summarized in Figure 8b. The highest content of M-O-M bonds of 58.98% was found in the GHTYZ film. The relatively low proportion of M-O-M bonds and the significant proportional variation among different combinations are postulated to be the results of the oxygen chemical states being strongly influenced by the complex internal structure of HEMOs; furthermore, larger metal–oxygen binding energy differences among elements may result in more oxygen vacancies [38]. To investigate this phenomenon, the average binding energy differences for six HEM combinations were calculated using reported values for individual metal–oxygen bonds [39–44]. Figure 8c reveals a correlation between the average binding energy differences and the proportion of M-O-M bonds in the films. The larger the average difference in binding energy of the metal–oxide bonds in the HEMOs films, the more complex the internal structure of the film is, increasing the proportion of non-well-bound oxygen. Based on the above analysis, the complexity of the internal structure of HEMO films can be increased by increasing the average difference of the binding energy, which suppresses the formation of conductive channels and lowers the device leakage current. However, excessive structural complexity can also lead to the formation of more oxygen-related defects, resulting in an increased leakage current.

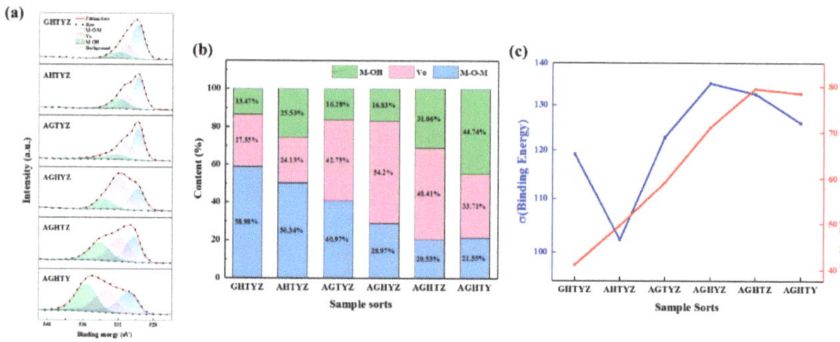

Figure 8. (**a**) XPSO1s spectra of HEMOs films; (**b**) percentage of three types of oxygen chemical states of HEMOs films; (**c**) the average difference in oxygen binding energy and the percentage of non-well-bound oxygen of HEMOs films.

The individual metal oxides were uniformly distributed in the film by XPS testing of the film. Moreover, the oxygen vacancies of the films were analysed. High entropy metal oxides usually have a low M-O-M content due to their complex structure. The complex structure hinders the formation of conductive channels and thus reduces the leakage current. However, the defects brought by the complex structure may also increase the leakage current. Therefore, it needs to be further analysed in conjunction with thin film leakage current testing.

3.3. Devices Properties

3.3.1. MIM Properties

The current-voltage (I-V) curves of the MIM devices and the leakage current density @1 MV/cm of the films are shown in Figure 9a,b. The elemental composition in the high entropy film exhibited a significant influence on the leakage current density of the film. AGHTZ and AGHTY films displayed large leakage current densities (>1 $\times$ 10^{-5} A/cm^2@1 MV/cm). These two groups of films have a rough surface as well as loose internal structure, with only approximately 20% of well-bonded oxygen content. The presence of a large number of oxygen-related defects primarily contributed to the increase in leakage current density. The AGTYZ film exhibited the lowest leakage current density (1.2 $\times$ 10^{-8} A/cm^2@1 MV/cm), which is conjectured to its larger average difference in binding energy in the film and lower percentage of non-well-bound oxygen.

In factor analysis, each element's impact was quantified by comparing the average property values between compositions with and without the specific element. The factor of each element and property was calculated as follows: To determine the impact of element X on property Y, the average value of property Y for all compositions containing X (designated as X = 1) was compared with the average value of property Y for all compositions without X (designated as X = 0). A factor value of X = 1 indicates that the element enhances the specific property under investigation, while a factor value of X = 0 suggests an inhibitory effect. This systematic analysis enables the isolation and quantification of each element's contribution to each property of HEMO films.

Factor analysis was performed to evaluate the impact of six metal oxide (AlO_x, GaO_x, HfO_x, TiO_x, YO_x, and ZrO_x) on the leakage current density in HEMOs films, as presented in Figure 9c. The presence of AlO_x and GaO_x was found to reduce the leakage current density. The introduction of Al helps to reduce the film roughness and the interface trap density, while the wide forbidden bandgap of Al_2O_3 suppresses the electron leap. Meanwhile, Ga_2O_3 has a large positive Gibbs free energy, which can effectively suppress the deterioration of the film dielectric properties caused by the hygroscopic reaction [6]. Hf and Ti addition significantly increased the leakage current density of HEMOs films, which can be attributed to the residual chloride ions, nitrate ions, and organicsintroduced in the precursor solutions ($HfCl_4$, $C_{16}H_{36}O_4Ti$, EGME) require higher thermal energy for decomposition and remain partially undecomposed after annealing, creating additional electron transport pathways in the dielectric films. However, the large positive Gibbs free energy of TiO_2 may promote film deterioration through hygroscopic reaction, while the narrow forbidden bandgap of TiO_2 facilitates electron leap in the films, increasing leakage current density in HEMOs films [45].

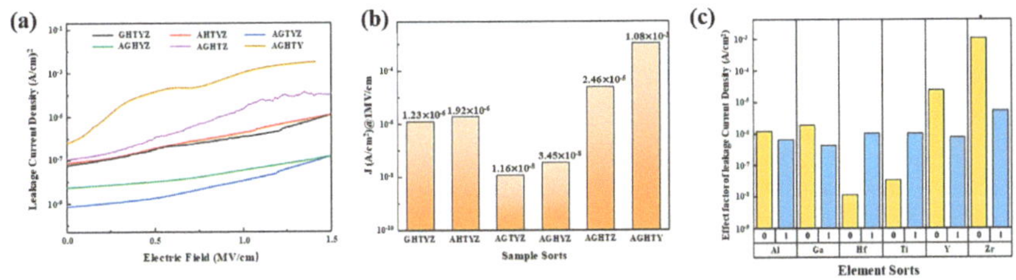

Figure 9. Film I-V characteristics: (**a**) I-V curves; (**b**) leakage current density under an electric field of 1 MV/cm; (**c**) factor analysis of each element's impact on leakage current density.

The capacitance-voltage (C-V) and capacitance-frequency (C-F) curves are shown in Figure 10a,b. Under a 1000 Hz electric field, the capacitance density and dielectric constant of HEMOs films were investigated, as summarized in Figure 10c. From the C-V curves, it can be observed that under a fixed-frequency electric field, the capacitance of the film is almostconstant with varying applied voltage, showing good voltage stability. The C-F measurements up to 1 MHz demonstrate that the capacitance decreases as frequency increases, suggesting a decrease in capacitance density and dielectric constant. This can be attributed to the polarization in the dielectric layer under an alternating electric field, where the dielectric constant is closely related to different types of polarization occurring at various frequencies within the film [46]. At low frequencies ($<10^2$ Hz), almost all polarization mechanisms can respond to the change in electric field, resulting in higher polarization and consequently larger dielectric constant. As the frequency increases, some polarization processes fail to keep pace with the field variations, decreasing polarization and then the dielectric constant [47].

The HEMOs films exhibit a highly disordered and asymmetric internal structure due to the incorporation of various smetallic elements with different atomic radii, enhancing atomic polarization under external electric field, contributing to higher dielectric constants. Among the studied films, AGHTZ film exhibits the largest capacitance density and dielectric constant at low frequencies, but undergoes rapid degradation with increasing frequency. The capacitance decreases directly from 4.89×10^{-10} to 1.28×10^{-12} at frequencies above 1000 Hz. As depicted in Figure 9b, the AGHTZ film exhibits a large leakage current density, possibly due to residual nitrates, hydroxides, or organic groups, as well as potential absorption of moisture from the air, increasing mobile charges. The presence of this mobile charges increase the overall polarization in the low-frequency range, but the capacitance-frequency stability decreases in the mid to high-frequency range. To better analyze the role of different elements in the HEMOs films on the dielectric properties, the effects of six metal oxide, AlO_x, GaO_x, HfO_x, TiO_x, YO_x, and ZrO_x, on the capacitance density and dielectric constant of the films were under factor analysis, as shown in Figure 10d,e. The result indicates that Ti significantly enhances capacitance density and dielectric constant, consistent with the high dielectric constant (60~80) in previous studies [16].

Considering all the above analysis, it leads to a conclusion that good surface properties result in excellent dielectric properties. YO_x and ZrO_x show good compatibility with other metal oxides, forming HEMOs film with good surface morphology. This superior quality also facilitates the formation of a dielectric HEMO layer and metal oxide semiconductors with low interface trap density. The wide bandgap of Al_2O_3 suppress the electron leaps within HEMOs film, while the large positive Gibbs free energy of Ga inhibits the degradation of dielectric properties caused by hygroscopic reaction. Additionally, the high crystallization temperature of Al_2O_3 and Ga_2O_3 enables the HEMOs film to maintain an amorphous state after 400 °C annealing, thereby reduces the formation of conductive channelsand the leakage current. Among 6 components, TiO_2 has the largest dielectric constant, significantly enhancing the dielectric constant of HEMOs films. The optimized films demonstrated excellent optical and electrical properties, including a high visible light transmission of 93.8%, low leakage current density of 1.2×10^{-8} A/cm^2@1 MV/cm, a high dielectric constant of 29.95 at 1000 Hz, and good frequency stability.

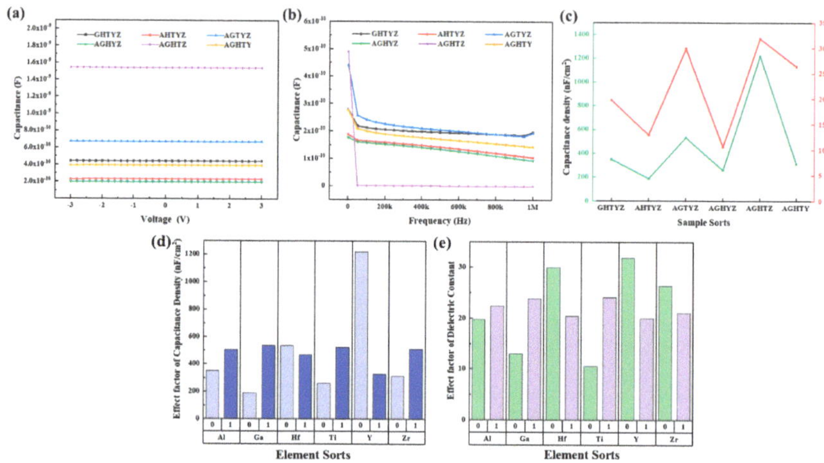

Figure 10. (**a**) C-V curves; (**b**) C-F curves; (**c**) Capacitance density, relative dielectric constant; (**d**) factor analysis of each element's impact on capacitance density; (**e**) factor analysis of each element's impact on dielectric constant.

Previous studies have shown that while single TiO_2 dielectric layers exhibit high dielectric constants (k: 60–80), their relatively low bandgap (3.45 eV) leads to large leakage currents, limiting their application as high-performance dielectric layer. However, the AGTYZ films containing Ti maintain both a high dielectric constant and a high optical bandgap of 5.26 eV. To further investigate the role of Ti in modulating the dielectric constant and optical bandgap, the Ti concentrations in AGTYZ films were varied, as detailed in Table 2.

Table 2. Ti concentration settings.

Elemental Components	The Concentration of Ti (M)	The Concentration of Remaining Elements (M)	Total Concentration (M)
AGTYZ	0	0.25	1.0
	0.12	0.22	
	0.20	0.20	
	0.28	0.18	

The optical and capacitive properties of HEMOs films with different concentrations of Ti were measured, as shown in Figure 11a–c. The optical bandgap of the films was fitted, the capacitance density and dielectric constant were calculated, as shown in Figure 11d–f. Results indicate that Ti concentration significantly modulates both optical band gap and dielectric properties of the films. Specifically, Ti concentration exhibits an inverse relationship with the optical band gap and a positive correlation with capacitance density and dielectric constant, confirming our previous factor analysis results. As the concentration of Ti increases from 0 to 0.28 M, the capacitance density of the films increases from 259.72 nF/cm^2 to 1133.50 nF/cm^2, and the dielectric constant rises significantly from 10.63 to 52.74, while the optical band gap shows a modest decrease from 5.62 eV to 5.05 eV. These findings suggest that adjusting Ti content in AGTYZ film enables dielectric HEMO layers to combine high dielectric constant with wide optical bandgap.

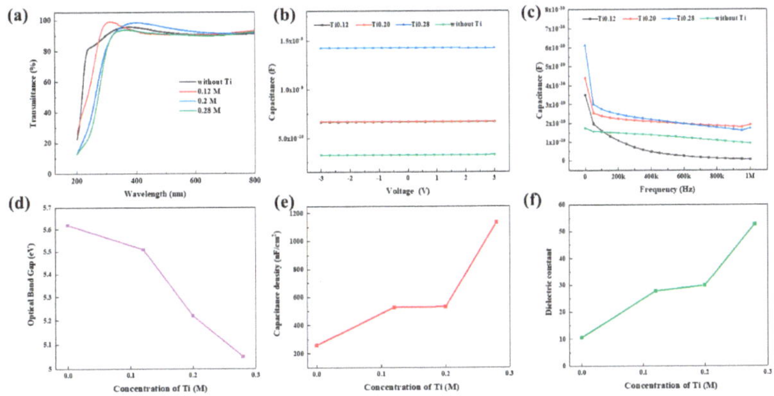

Figure 11. Effect of Ti concentration on film properties: (**a**) transmittance; (**b**) C-V curves; (**c**) C-F curves; (**d–f**) Plot of Ti concentration versus optical band gap, capacitance density, dielectric constant.

By comparing the performance of MIM devices prepared from six metal oxides, including leakage current and dielectric constant, we found that AGTYZ has the best performance. And in the factor analysis, Ti has the effect of increasing the dielectric constant, so we investigate the effect of Ti content in AGTYZ on the film properties. The dielectric constant of the final dielectric film increased to 52.74.

3.3.2. TFT Properties

Given their excellent surface morphology and dielectric properties, TFT devices with a magnetron-sputtered IGZO active layer and an AGTYZ dielectric layer were prepared. The device characteristics are shown in Figure 12a,b. The TFTs exhibited a switching ratio of 1.0×10^5, a mobility of 18.2 cm^2/Vs, am SS of 0.288 V/decade, and a V_{th} of -0.203 V. More importantly, they exhibited a minimal hysteresis effect. In general, the cause of the hysteresis effect is theorized to be carrier trapping at defect sites within the active layer or at interfaces between layers [48]. These results demonstrate that AGTYZ films possess low defect density and high-quality interfaces between the active and dielectric layers. As shown in Table 3, in contrast to other TFTs prepared with single-component dielectric layers, the TFT with a dielectric AGTYZ layer investigated in this study exhibits higher mobility, a higher switching ratio, a lower operating voltage, and subthreshold swing. By combining the advantageous properties of Al_2O_3, Ga_2O_3, TiO_2, Y_2O_3, and ZrO_2, high-performance dielectric HEMO layers were successfully developed and implemented in thin-film transistors.

Table 3. Comparison between AGTYZ TFTs and other single component TFTs.

Dielectric	Ref.	Mobility (cm^2/Vs)	I_{on}/I_{off}	Vth (V)	SS (V/dec)
AGTYZ		18.2	10^5	−0.203	0.288
Al	[49]	4.7	10^3	0.2	0.24
Ga	[50]	3.09	10^5	0.83	0.18
Ti	[51]	0.2	10^3	−24	/
Y	[52]	34	10^5	−4	/
Zr	[53]	12.7	10^5	0.6	0.36

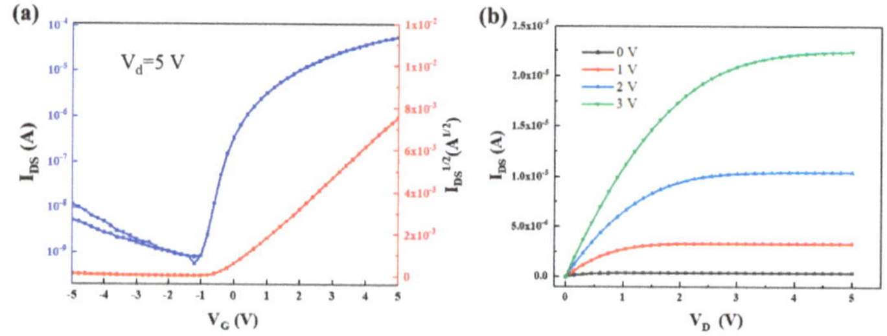

Figure 12. TFT performance characterization: (**a**) TFT transfer curve and and $I_{DS}^{1/2}$-V_G curve; (**b**) TFT output curve.

4. Conclusions

In this paper, five-element HEMO films were prepared by spin coating, and the effects of the components on the properties of HEMO films were investigated. The results show that $(AlGaTiYZr)O_x$ film exhibits the best dielectric properties among the six investigated five-element HEMOs, with more than 90% visible transmittance, an optical bandgap of 5.26 eV, a leakage current density of 1.2×10^{-8} A/cm^2 at a field strength of 1 MV/cm, and a high dielectric constant of 29.95 under a 1000 Hz electric field. On this basis, we prepared IGZO TFTs with dielectric HEMO layers to investigate the suitability of HEMO films for TFTs. The test results show that the TFTs based on a dielectric $(AlGaTiYZr)O_x$ layer exhibit a mobility of 18.2 cm^2/Vs, a V_{th} of -0.203 V, and an SS of 0.288 V/decade, along with a minimal hysteresis, suggesting the potential of applying HEMOs to TFTs. In addition, we factorized the performance parameters of the HEMO films and found that Ti has a significant enhancing effect on the dielectric constant. Based on this, we achieved the modulation of the dielectric constant by adjusting the ratio of Ti in the HEMO films, increasing the dielectric constant to 52.74 and maintaining a large optical bandgap of 5.05 eV. Overall, dielectric HEMO films prepared by spin coating can synthesize the advantages of multiple high-k dielectrics to obtain better film properties while having the advantages of simple processing, low-temperature preparation, and low cost. HEMO films are expected to be widely used as dielectric layers in the future for flexible, transparent, and high-performance electronic devices.

Author Contributions: Conceptualization: J.L. and X.X. Methodology: H.L., Z.L. and H.N. Investigation: X.H. Resources: R.Y., Y.S. and X.W. Writing—original draft: J.L. and X.X. Writing—review and editing: Y.S., J.L. and X.X. Visualization: X.W. Supervision: Y.W. All authors have read and agreed to the published version of the manuscript.

Funding: This work was supported by the Hunan Provincial Natural Science Foundation of China (2023JJ30499), the Guangdong Provincial Natural Science Foundation of China (2022A1515010198), the Southwest Institute of Technology and Engineering Cooperation Fund (HDHDW59A020301). This project was also supported by the Scientific Research Fund of the Hunan Provincial Education Department, China (No. 22A0289).

Data Availability Statement: The raw data supporting the conclusions of this article will be made available by the authors on request.

Conflicts of Interest: The authors declare no conflicts of interest.

References

1. Ning, H.L.; Liang, Z.H.; Fu, X.; Yao, R.H.; Xu, Z.H.; Qiu, T.; Yang, Z.; Hu, C.Y.; Xu, W.; Peng, J.B. Environmentally friendly, flexible and high performance PVA dielectric layer fabricated by solution method and its application in IGZO-TFT. *Org. Electron.* **2022**, *100*, 106383. [CrossRef]
2. Kol, S.; Oral, A.Y. Hf-Based High-κ Dielectrics: A Review. *Acta Phys. Pol. A* **2019**, *136*, 873–881. [CrossRef]

3. Zhou, S.X.; Zhang, J.H.; Guo, X.P.; Ning, H.L.; Guo, D.; Yao, R.H.; Zhu, Z.N.; Liang, Z.H.; Fang, Z.Q.; Peng, J.B. Inkjet printing of homogeneous and green cellulose nanofibril dielectrics for high performance IGZO TFTs. *J. Mater. Chem. C* **2020**, *8*, 12578–12586. [CrossRef]
4. Robertson, J. High dielectric constant gate oxides for metal oxide Si transistors. *Rep. Prog. Phys.* **2006**, *69*, 327–396. [CrossRef]
5. Newsom, T.L.; Allemang, C.R.; Cho, T.H.; Dasgupta, N.P.; Peterson, R.L. 59.9 $mV{\cdot}dec^{-1}$ Subthreshold Swing Achieved in Zinc Tin Oxide TFTs With In Situ Atomic Layer Deposited Al_2O_3 Gate Insulator. *IEEE Electron Device Lett.* **2023**, *44*, 72–75. [CrossRef]
6. Xu, W.Y.; Cao, H.T.; Liang, L.Y.; Xu, J.B. Aqueous Solution-Deposited Gallium Oxide Dielectric for Low-Temperature, Low-Operating-Voltage Indium Oxide Thin-Film Transistors: A Facile Route to Green Oxide Electronics. *ACS Appl. Mater. Interfaces* **2015**, *7*, 14720–14725. [CrossRef]
7. Zhang, Z.H.; Su, G.P.; Ning, H.L.; Jiang, B.C.; Liu, Y.X.; Xiong, X.; Jiang, Y.S.; Liu, D.R.; Yao, R.H.; Peng, J.B. Solution-processed transparent PVP:HfO_2 hybrid dielectric films with low leakage current density and high k. *Surf. Interfaces* **2023**, *42*, 10. [CrossRef]
8. Li, Y.; Zhu, L.; Chen, C.S.; Zhu, Y.; Wan, C.J.; Wan, Q. High-Performance Indium-Gallium-Zinc-Oxide Thin-Film Transistors with Stacked Al_2O_3/HfO_2 Dielectrics. *Chin. Phys. Lett.* **2022**, *39*, 118501. [CrossRef]
9. Yang, Y.; Xu, Z.; Qiu, T.; Ning, H.; Zhong, J.; Li, M.; Luo, D.; Liu, X.; Yao, R.; Peng, J. High k PVP titanium dioxide composite dielectric with low leakage current for thin film transistor. *Org. Electron.* **2022**, *101*, 106413. [CrossRef]
10. Nirosha, R.; Agarwal, R. Gate dielectric based steady state & transient analysis of channel characteristics for organic thin-film transistors. *J. Mater. Sci.-Mater. Electron.* **2023**, *34*, 2120. [CrossRef]
11. Pan, T.M.; Chang, C.J. High-performance poly-silicon TFTs with high-k Y_2O_3 gate dielectrics. *Semicond. Sci. Technol.* **2011**, *26*, 075004. [CrossRef]
12. Wang, Y.W.; Wang, S.; Ye, H.D.; Zhang, W.H.; Xiang, L. Negative Bias Temperature Instability in Top-Gated Carbon Nanotube Thin Film Transistors With Y_2O_3 Gate Dielectric. *IEEE Trans. Device Mater. Reliab.* **2023**, *23*, 571–576. [CrossRef]
13. Hasan, M.M.; Roy, S.; Tokumitsu, E.; Chu, H.Y.; Kim, S.C.; Jang, J. High performance, amorphous InGaZnO thin-film transistors with ferroelectric ZrO_2 gate insulator by one step annealing. *Appl. Surf. Sci.* **2023**, *611*, 155533. [CrossRef]
14. Salunkhe, P.; Kekuda, D. p-channel NiO thin film transistors grown with high k ZrO_2 gate oxide for low voltage operation. *Phys. Scr.* **2023**, *98*, 065913. [CrossRef]
15. Zhu, C.D.; Liu, A.; Liu, G.X.; Jiang, G.X.; Meng, Y.; Fortunato, E.; Martins, R.; Shan, F.K. Low-temperature, nontoxic water-induced high-k zirconium oxide dielectrics for low-voltage, high-performance oxide thin-film transistors. *J. Mater. Chem. C* **2016**, *4*, 10715–10721. [CrossRef]
16. Verma, A.; Mishra, V.N.; Prakash, R. Self-Assembled Au/P3HT, High-k Bilayer Dielectric-Based Solution Processed Low Voltage OTFT for Multiparametric Ammonia Sensor at Room Temperature. *IEEE Trans. Electron Devices* **2023**, *70*, 281–287. [CrossRef]
17. Jakschik, S.; Schroeder, U.; Hecht, T.; Gutsche, M.; Seidl, H.; Bartha, J.W. Crystallization behavior of thin ALD-Al_2O_3 films. *Thin Solid Film.* **2003**, *425*, 216–220. [CrossRef]
18. Lee, S.H.; Bak, S.Y.; Park, C.Y.; Baek, D.; Yi, M.S. Enhancement of electrical performance in indium-zinc oxide thin-film transistors with HfO_2/Al_2O_3 gate insulator deposited via low-temperature ALD. *Displays* **2023**, *80*, 102566. [CrossRef]
19. Song, X.J.; Xu, J.P.; Liu, L.; Deng, Y.H.; Lai, P.T.; Tang, W.M. Optimizing Al-doped ZrO_2 as the gate dielectric for MoS_2 field-effect transistors. *Nanotechnology* **2020**, *31*, 135206. [CrossRef] [PubMed]
20. Zhou, Y.; Liang, Z.H.; Yao, R.H.; Zuo, W.C.; Zhou, S.X.; Zhu, Z.N.; Wang, Y.P.; Qiu, T.; Ning, H.L.; Peng, J.B. Effect of Zirconium Doping on Electrical Properties of Aluminum Oxide Dielectric Layer by Spin Coating Method with Low Temperature Preparation. *Coatings* **2020**, *10*, 620. [CrossRef]
21. Peng, J.B.; Wei, J.L.; Zhu, Z.N.; Ning, H.L.; Cai, W.; Lu, K.K.; Yao, R.H.; Tao, H.; Zheng, Y.Q.; Lu, X.B. Properties-Adjustable Alumina-Zirconia Nanolaminate Dielectric Fabricated by Spin-Coating. *Nanomaterials* **2017**, *7*, 419. [CrossRef] [PubMed]
22. Ma, P.F.; Sun, J.M.; Zhang, G.Q.; Liang, G.D.; Xin, Q.; Li, Y.X.; Song, A.M. Low-temperature fabrication of HfAlO alloy dielectric using atomic-layer deposition and its application in a low-power device. *J. Alloy. Compd.* **2019**, *792*, 543–549. [CrossRef]
23. Yang, H.Y.; Liang, Z.H.; Fu, X.; Xu, Z.H.; Ning, H.L.; Liu, X.Z.; Lin, J.J.; Pan, Y.R.; Yao, R.H.; Peng, J.B. Application of Amorphous Zirconium-Yttrium-Aluminum-Magnesium-Oxide Thin Film with a High Relative Dielectric Constant Prepared by Spin-Coating. *Membranes* **2021**, *11*, 608. [CrossRef] [PubMed]
24. Li, H.Y.; Zhou, Y.; Liang, Z.H.; Ning, H.L.; Fu, X.; Xu, Z.H.; Qiu, T.; Xu, W.; Yao, R.H.; Peng, J.B. High-Entropy Oxides: Advanced Research on Electrical Properties. *Coatings* **2021**, *11*, 628. [CrossRef]
25. Zhang, Y.; Zuo, T.T.; Tang, Z.; Gao, M.C.; Dahmen, K.A.; Liaw, P.K.; Lu, Z.P. Microstructures and properties of high-entropy alloys. *Prog. Mater. Sci.* **2014**, *61*, 1–93. [CrossRef]
26. Salian, A.; Pujar, P.; Vardhan, R.V.; Cho, H.W.; Kim, S.; Mandal, S. Evolution of High Dielectric Permittivity in Low-Temperature Solution Combustion-Processed Phase-Pure High Entropy Oxide (CoMnNiFeCr)O for Thin Film Transistors. *ACS Appl. Electron. Mater.* **2023**, *5*, 2608–2623. [CrossRef]
27. Liang, Z.H.; Wu, W.J.; Fu, X.; Ning, H.L.; Xu, W.; Xiong, X.; Qiu, T.; Luo, C.; Yao, R.H.; Peng, J.B. Flexible High-Entropy Poly(vinyl alcohol) Dielectric Films Were Prepared at a Low Temperature and Applied to an Indium Gallium Zinc Oxide Thin-Film Transistor. *J. Phys. Chem. Lett.* **2023**, *14*, 9245–9249. [CrossRef]
28. Acharya, V.; Agarwal, K.; Mondal, S. Electronic materials for solution-processed TFTs. *Mater. Res. Express* **2023**, *10*, 082002. [CrossRef]

29. Vukmirovic, J.; Tripkovic, D.; Bajac, B.; Kojic, S.; Stojanovic, G.M.; Srdic, V.V. Comparison of barium titanate thin films prepared by inkjet printing and spin coating. *Process. Appl. Ceram.* **2015**, *9*, 151–156. [CrossRef]
30. Sahoo, S.; Arora, A.; Doshi, P. Two-layer spin coating flow of Newtonian liquids: A computational study. *Comput. Fluids* **2016**, *131*, 180–189. [CrossRef]
31. Liang, Z.; Zhou, S.; Cai, W.; Fu, X.; Ning, H.; Chen, J.; Yuan, W.; Zhu, Z.; Yao, R.; Peng, J. Zirconium-aluminum-oxide dielectric layer with high dielectric and relatively low leakage prepared by spin-coating and the application in thin-film transistor. *Coatings* **2020**, *10*, 282. [CrossRef]
32. Boratto, M.H.; Congiu, M.; dos Santos, S.B.; Scalvi, L.V. Annealing temperature influence on sol-gel processed zirconium oxide thin films for electronic applications. *Ceram. Int.* **2018**, *44*, 10790–10796. [CrossRef]
33. Zhou, S.; Zhang, J.; Fang, Z.; Ning, H.; Cai, W.; Zhu, Z.; Liang, Z.; Yao, R.; Guo, D.; Peng, J. Thermal effect of annealing-temperature on solution-processed high-k ZrO 2 dielectrics. *RSC Adv.* **2019**, *9*, 42415–42422. [CrossRef] [PubMed]
34. Hu, B.; Yao, M.; Yang, P.; Chen, J.; Yao, X. Preparation of Al_2O_3 dense films using sol derived from $Al(NO_3)_3{\cdot}9H_2O$. *Mater. Technol.* **2014**, *29*, 47–51. [CrossRef]
35. Hu, B.; Jia, E.; Du, B.; Yin, Y. A new sol-gel route to prepare dense Al_2O_3 thin films. *Ceram. Int.* **2016**, *42*, 16867–16871. [CrossRef]
36. Koslowski, N.; Hoffmann, R.C.; Trouillet, V.; Bruns, M.; Foro, S.; Schneider, J.J. Synthesis, oxide formation, properties and thin film transistor properties of yttrium and aluminium oxide thin films employing a molecular-based precursor route. *Rsc Adv.* **2019**, *9*, 31386–31397. [CrossRef] [PubMed]
37. Li, Y.J.; Xu, W.Y.; Liu, W.J.; Han, S.; Cao, P.J.; Fang, M.; Zhu, D.L.; Lu, Y.M. High-Performance Thin-Film Transistors with Aqueous Solution-Processed NiInO Channel Layer. *ACS Appl. Electron. Mater.* **2019**, *1*, 1842–1851. [CrossRef]
38. Liang, Z.H.; Wu, W.J.; Fu, X.; Ning, H.L.; Su, G.P.; Wang, H.C.; Qiu, T.; Yang, Z.; Yao, R.H.; Peng, J.B. Solution-processed high entropy metal oxides as dielectric layers with high transmittance and performance and application in thin film transistors. *Surf. Interfaces* **2023**, *40*, 103147. [CrossRef]
39. Gobrecht, D.; Plane, J.M.C.; Bromley, S.T.; Decin, L.; Cristallo, S.; Sekaran, S. Bottom-up dust nucleation theory in oxygen-rich evolved stars. *Astron. Astrophys.* **2022**, *658*, A167. [CrossRef]
40. Gowtham, S.; Costales, A.; Pandey, R. Theoretical study of neutral and ionic states of small clusters of Ga_mO_n (m, n = 1, 2). *J. Phys. Chem. B* **2004**, *108*, 17295–17300. [CrossRef]
41. Ariyarathna, I.R.; Duan, C.R.; Kulik, H.J. Understanding the chemical bonding of ground and excited states of HfO and HfB with correlated wavefunction theory and density functional approximations. *J. Chem. Phys.* **2022**, *156*, 184113. [CrossRef] [PubMed]
42. Kushawaha, V.S. Potential energy curves and dissociation energy of titanium monoxide. *J. Phys. Chem.* **1973**, *77*, 2885–2888. [CrossRef]
43. Sievers, M.R.; Armentrout, P.B. Oxidation of CO and reduction of CO_2 by gas phase Zr^+, ZrO^+, and $ZrO_2{}^+$. *Int. J. Mass Spectrom.* **1999**, *185*, 117–129. [CrossRef]
44. Jing-Feng, L.; Watanabe, R.; Bo-Ping, Z.; Asami, K.; Hashimoto, K. X-ray photoelectron spectroscopy investigation on the low-temperature degradation of 2 mol% Y_2O_3-ZrO_2 ceramics. *J. Am. Ceram. Soc.* **1996**, *79*, 3109–3112. [CrossRef]
45. Zhao, Y.; Kita, K.; Toriumi, A. Thermodynamic analysis of moisture absorption phenomena in high-permittivity oxides as gate dielectrics of advanced complementary-metal-oxide-semiconductor devices. *Appl. Phys. Lett.* **2010**, *96*, 242901. [CrossRef]
46. Wang, S.; Yang, C.; Li, X.M.; Jia, H.Y.; Liu, S.R.; Liu, X.Y.; Minari, T.; Sun, Q.Q. Polymer-based dielectrics with high permittivity and low dielectric loss for flexible electronics. *J. Mater. Chem. C* **2022**, *10*, 6196–6221. [CrossRef]
47. Qiu, J.; Gu, Q.; Sha, Y.; Huang, Y.; Zhang, M.; Luo, Z.Y. Preparation and application of dielectric polymers with high permittivity and low energy loss: A mini review. *J. Appl. Polym. Sci.* **2022**, *139*, e52367. [CrossRef]
48. Chang, Y.; Bukke, R.N.; Bae, J.; Jang, J. Low-Temperature Solution-Processed HfZrO Gate Insulator for High-Performance of Flexible LaZnO Thin-Film Transistor. *Nanomaterials* **2023**, *13*, 2410. [CrossRef]
49. Xia, D.; Xu, J. High mobility and low operating voltage ZnGaO and ZnGaLiO transistors with spin-coated Al_2O_3 as gate dielectric. *J. Phys. D Appl. Phys.* **2010**, *43*, 442001. [CrossRef]
50. Chen, L.; Xu, W.; Liu, W.; Han, S.; Cao, P.; Fang, M.; Zhu, D.; Lu, Y. Polymer-assisted deposition of gallium oxide for thin-film transistor applications. *ACS Appl. Mater. Interfaces* **2019**, *11*, 29078–29085. [CrossRef] [PubMed]
51. Cai, Q.J.; Gan, Y.; Chan-Park, M.B.; Yang, H.B.; Lu, Z.S.; Song, Q.L.; Li, C.M.; Li Dong, Z. Solution-processable organic-capped titanium oxide nanoparticle dielectrics for organic thin-film transistors. *Appl. Phys. Lett.* **2008**, *93*, 113304. [CrossRef]
52. Adamopoulos, G.; Thomas, S.; Bradley, D.D.; McLachlan, M.A.; Anthopoulos, T.D. Low-voltage ZnO thin-film transistors based on Y_2O_3 and Al_2O_3 high-k dielectrics deposited by spray pyrolysis in air. *Appl. Phys. Lett.* **2011**, *98*, 123503. [CrossRef]
53. Zhou, S.; Fang, Z.; Ning, H.; Cai, W.; Zhu, Z.; Wei, J.; Lu, X.; Yuan, W.; Yao, R.; Peng, J. Bias stability enhancement in thin-film transistor with a solution-processed ZrO_2 dielectric as gate insulator. *Appl. Sci.* **2018**, *8*, 806. [CrossRef]

Disclaimer/Publisher's Note: The statements, opinions and data contained in all publications are solely those of the individual author(s) and contributor(s) and not of MDPI and/or the editor(s). MDPI and/or the editor(s) disclaim responsibility for any injury to people or property resulting from any ideas, methods, instructions or products referred to in the content.

micromachines

Article

Annealing Study on Praseodymium-Doped Indium Zinc Oxide Thin-Film Transistors and Fabrication of Flexible Devices

Zhenyu Wu [1], Honglong Ning [1,2], Han Li [3], Xiaoqin Wei [3], Dongxiang Luo [4,5,6,*], Dong Yuan [5], Zhihao Liang [1], Guoping Su [1], Rihui Yao [1,7,*] and Junbiao Peng [1]

1 Guangdong Basic Research Center of Excellence for Energy & Information Polymer Materials, State Key Laboratory of Luminescent Materials and Devices, School of Materials Sciences and Engineering, South China University of Technology, Guangzhou 510640, China; ninghl@scut.edu.cn (H.N.)
2 The International School of Microelectronics, Dongguan University of Technology, Dongguan 523808, China
3 Southwest Institute of Technology and Engineering, Chongqing 400039, China
4 Huangpu Hydrogen Innovation Center/Guangzhou Key Laboratory for Clean Energy and Materials, School of Chemistry and Chemical Engineering, Guangzhou University, Guangzhou 510006, China
5 Guangdong Provincial Key Laboratory of Optical Information Materials and Technology, South China Academy of Advanced Optoelectronics, South China Normal University, Guangzhou 510006, China
6 Key Laboratory of Optoelectronic Devices and Systems of Ministry of Education and Guangdong Province, College of Physics and Optoelectronic Engineering, Shenzhen University, Shenzhen 518060, China
7 Key Lab of Guangdong Province for High Property and Functional Polymer Materials, South China University of Technology, Guangzhou 510640, China
* Correspondence: luodx@gzhu.edu.cn (D.L.); yaorihui@scut.edu.cn (R.Y.)

Abstract: The praseodymium-doped indium zinc oxide (PrIZO) thin-film transistor (TFT) is promising for applications in flat-panel displays, due to its high carrier mobility and stability. Nevertheless, there are few studies on the mechanism of annealing on PrIZO films and the fabrication of flexible devices. In this work, we first optimized the annealing-process parameters on the glass substrate. As the annealing temperature rises, the film tends to be denser and obtains a lower surface roughness, a narrower optical-band gap and less oxygen-vacancy content. However, the μ-PCD test shows the 250 °C-annealed film obtains the least defects. And the PrIZO TFT annealed at 250 °C exhibited a desired performance with a saturation mobility (μ_{sat}) of 14.26 $cm^2 \cdot V^{-1} \cdot s^{-1}$, a subthreshold swing (SS) of 0.14 $V \cdot dec^{-1}$, an interface trap density (D_{it}) of 3.17×10^{11}, an I_{on}/I_{off} ratio of 1.83×10^8 and a threshold voltage (Vth) of −1.15 V. The flexible devices were prepared using the optimized parameters on the Polyimide (PI) substrate and subjected to static bending tests. After bending at a radius of 5 mm, the mobility of devices decreases slightly from 12.48 to 10.87 $cm^2 \cdot V^{-1} \cdot s^{-1}$, demonstrating the great potential of PrIZO for flexible displays.

Keywords: praseodymium-doped; indium zinc oxide; annealing temperature; flexible; thin-film transistor

check for updates

Academic Editor: Chengyuan Dong

Received: 6 December 2024
Revised: 24 December 2024
Accepted: 25 December 2024
Published: 26 December 2024

Citation: Wu, Z.; Ning, H.; Li, H.; Wei, X.; Luo, D.; Yuan, D.; Liang, Z.; Su, G.; Yao, R.; Peng, J. Annealing Study on Praseodymium-Doped Indium Zinc Oxide Thin-Film Transistors and Fabrication of Flexible Devices. *Micromachines* **2025**, *16*, 17. https://doi.org/10.3390/mi16010017

Copyright: © 2024 by the authors. Licensee MDPI, Basel, Switzerland. This article is an open access article distributed under the terms and conditions of the Creative Commons Attribution (CC BY) license (https://creativecommons.org/licenses/by/4.0/).

1. Introduction

The thin-film transistor (TFT) for active-matrix organic light-emitting diode (AMOLED) has been widely studied with the rapid development of high-resolution, high refresh rate, and flexible and transparent-display technology [1–5]. The amorphous oxide semiconductor (AOS) material, such as indium gallium zinc oxide (IGZO) and indium zinc oxide (IZO), has gradually replaced traditional hydrogenated amorphous silicon (a-Si: H) and low-temperature polycrystalline silicon (LTPS), to be considered as the channel of TFT, due to its good uniformity, high mobility, high optical transparency and low off-current [6–11]. Nevertheless, AOS TFT suffers the problem of being sensitive to external environmental factors such as light, water and oxygen [12,13]. In order to address the problem of instability, the method of adding various dopants which have strong binding energy with oxygen to AOS has attracted people's attention [14–16]. For example, the praseodymium-doped indium zinc oxide (PrIZO) TFT, where Pr effectively suppress the formation of oxygen vacancies (V_O), exhibit obvious stability under negative-bias illumination stress (NBIS) [17–20].

Furthermore, during the film preparation process, a significant number of defect states are often generated [21,22], which adversely impacts the quality of the film and subsequently diminishes device performance [23]. Therefore, it is essential to implement a certain degree of post-treatment on the film, to enhance its quality. The most prevalent method for post-treatment is annealing. Numerous studies have demonstrated that employing thermal annealing can improve device performance by allowing atoms within the film to gain energy for rearrangement and facilitating the formation of oxide lattices [24,25]. Also, annealing effectively eliminates weak bonds, reduces defects and decrease the density of subgap state, which affect the stability of devices [26–30]. In this work, the annealing process at different temperatures are applied to the films and TFTs of PrIZO to explore the internal mechanism. It is discovered that annealing at appropriate temperature can effectively diminish the deep-level defects and enhance the carrier concentration. Based on the analysis of mechanism, the optimized annealing process is acquired and then applied to the preparation of flexible TFTs, which exhibit great performance.

2. Materials and Methods

The PrIZO films were prepared on an alkali-free glass (1.0×1.0 cm^2) by radio frequency (RF) magnetron sputtering using a PrIZO target (Pr:In:Zn = 0.2:5.2:1 at%). The sputtering power, chamber pressure and sputtering time were maintained at 80 W, 0.9 Pa and 6 min. The deposition process was executed with a gas mixing ratio of Ar:O_2 = 9:1. Then, the films were annealed at 200 °C, 250 °C and 300 °C in air for 30 min, respectively.

The schematic diagram of the PrIZO TFT prepared on the glass substrate is shown in Figure 1a. A 300 nm thick Al film was deposited by direct current (DC) sputtering and then patterned by wet etching to form the gate electrode. Then, part of Al film was oxidized to AlO_x as an insulating layer by anodic oxidation in an electrolyte consisting of ammonium tartrate (3.48 wt. %) and ethylene glycol (96.52 wt. %) at room temperature. The thickness of the insulating layer was about 200 nm and the unit-area capacitance of the insulating layer was about 38 nF/cm^2. As mentioned above, after sputtering a patterned layer of PrIZO, the as-deposited samples were annealed at a different temperature in air for 30 min. Finally, a 150 nm thick Al layer was prepared by DC sputtering through a shadow mask (channel width/length = 530 μm/390 μm) as the source/drain electrodes.

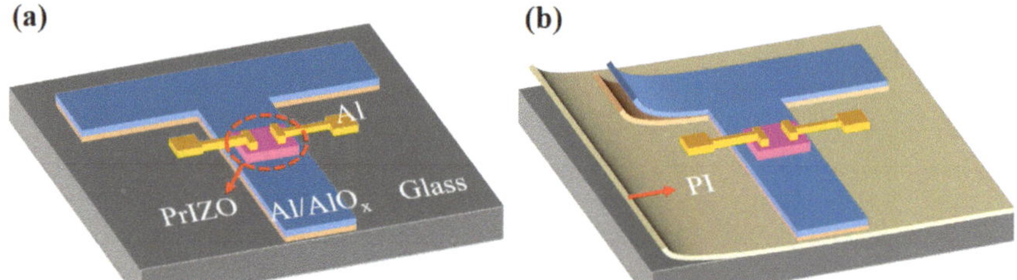

Figure 1. Schematic diagram of PrIZO TFT: (**a**) glass substrate, (**b**) PI substrate.

As shown in Figure 1b, the flexible PrIZO TFT device adopts a PI substrate (1.0×1.0 cm^2). The PI substrate is coated with a PI film on the glass. In addition, the structure of the device above the substrate is identical.

All sputtering operations are performed by a magnetron sputtering machine. (JGP-560, Sky Technology Development Co., Ltd., Shenyang, Chinese Academy of Sciences). The phase, thickness and density of PrIZO films were gained from an X-ray diffraction (XRD) and X-ray reflection (XRR) analyzer (Empyrean Nano edition, PANalytical, Almelo, The Netherlands). The surface morphology was observed by atomic force microscopy (AFM). The optical properties of PrIZO thin films were characterized by a UV-VIS spectrophotometer (Shimadzu UV-3600, Kyoto, Japan). The relative carrier concentrations of the films were determined by microwave photoconductivity decay (μ-PCD, LTA-1620SP, Kobelco, Kobe, Japan). The chemical changes of the PrIZO films were detected by X-ray photoelectron spectroscopy (XPS) measurements. The electrical performance of the device was measured by using a semiconductor parameter analyzer (FS-Pro-Semiconductor parameter tester, Primarius Technologies Co., Ltd., Shanghai, China) in dark and air environment.

3. Results and Discussion

3.1. XRD Analysis

As shown in Figure 2, it is evident that all the XRD patterns of films have a broad peak between 20° and 35°, which is caused by the glass substrate, and there are no other characteristic peaks beyond that. This indicates that all films remain amorphous under the treatment process in this study, which is related to the lattice difference of cubic In_2O_3, hexagonal ZnO and hexagonal Pr_2O_3 [31].

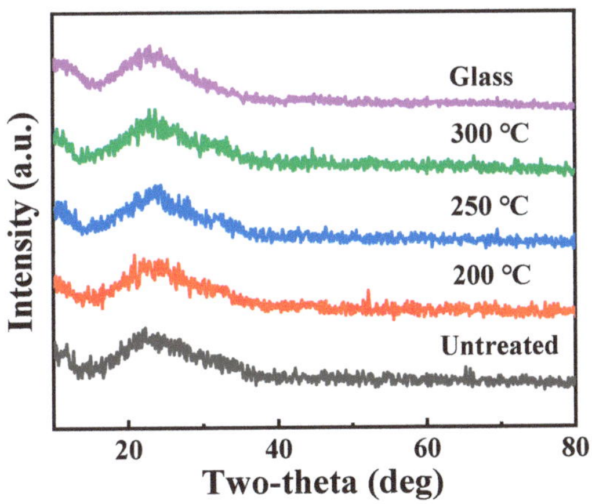

Figure 2. XRD patterns of PrIZO thin films annealed at different temperatures.

3.2. XRR Analysis

Figure 3a shows the XRR test results of PrIZO films annealed at different temperatures. The black curve is the data actually measured, and the red curve is the fitting curve. The two curves basically coincide, indicating that the fitting results are credible. Figure 3b shows the effect of annealing temperature on film thickness and density. From Figure 3b, the thickness and the density of untreated PrIZO thin films are about 16.1 nm and 6.25 g/cm^3. With the rise of annealing temperature, the thickness of the film decreases markedly, and the density increases accordingly. The thickness of the film annealed at 300 °C drops to approximately 15.2 nm, and the density of that rises to 6.42 g/cm^3. The rationale behind this phenomenon is that annealing supplies energy to the atoms or molecules of the film by means of heat transfer, allowing them to rearrange themselves.

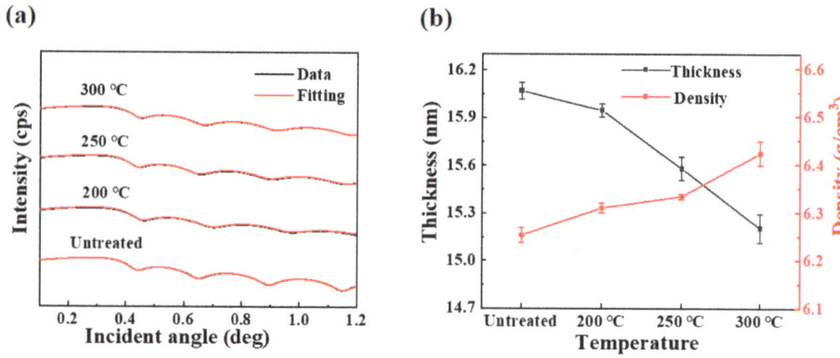

Figure 3. PrIZO thin films annealed at different temperature: (**a**) XRR, (**b**) thickness and density.

3.3. AFM Analysis

The 3D surface morphology of the films annealed at different temperature is shown in Figure 4a–d. As is apparent, the surface of the untreated film is uneven and contains numerous minor protuberances. As the annealing temperature increases, the protrusion gradually decreases, the surface tends to be smooth, and the surface roughness also reduces. It should be noticed that the surface morphology of the film did not change significantly until the temperature reached 250 °C. The morphology of film annealed at 300 °C is improved significantly, with a low roughness of 0.72 nm. This also implies that the molecular motion is relatively less intense at low temperatures, thereby causing only minor changes in roughness. Figure 4e,f show the surface morphology of the insulating layer on different substrates. It can be inferred that, because of an extra layer of PI film, the insulating layer on the PI substrate has a rougher interface. Moreover, it is worth mentioning that lower surface roughness could reduce carrier scattering and favorably increase the mobility of devices [32].

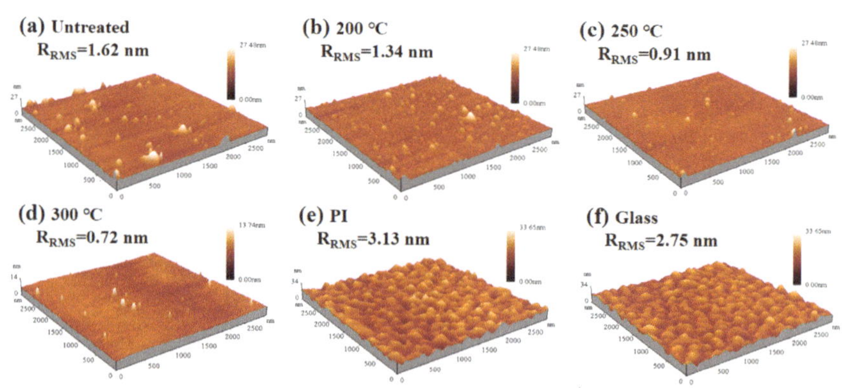

Figure 4. AFM images of PrIZO thin films: (**a**) untreated, (**b**) 200 °C, (**c**) 250 °C, (**d**) 300 °C and AFM images of insulators: (**e**) on the PI substrate, (**f**) on the glass substrate.

3.4. Optical Characterization

Figure 5a shows the transmission spectra of PrIZO films annealed at different temperature. The average transmittance of PrIZO films in the visible band exceeds 94%, which fully demonstrates the great potential application of PrIZO in transparent display. In addition, the optical band gap (E_g) of PrIZO films can be fitted from the absorption spectra, as shown in Figure 5b, according to Equation (1):

$$(\alpha h\nu)^2 = A(h\nu - Eg) \quad (1)$$

where α represents the absorption coefficient, $h\nu$ denotes the energy of the photon, and A constitutes a constant [33].

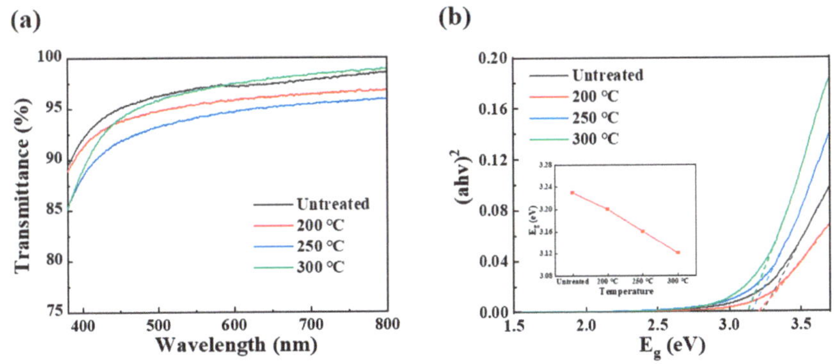

Figure 5. PrIZO thin films: (**a**) transmission spectra, (**b**) optical band gap.

As can be seen from Figure 5b, with the annealing temperature increasing to 300 °C, the band gaps of PrIZO films are, respectively, 3.23 eV, 3.20 eV, 3.16 eV and 3.12 eV. This indicates that increasing the annealing temperature effectively reduces the E_g of the PrIZO film, so that electrons can be excited more conveniently from the valence band to the conduction band, which is conducive to charge transport.

3.5. XPS Analysis

In PrIZO thin films, the XPS test is employed to analyze the changes in the surface composition and the various chemical states of oxygen, to further explain the mechanism of annealing treatment on TFT performance. Figure 6 shows the O 1s peaks of XPS spectra of PrIZO films annealed at different temperature. The O 1s peak can be fitted by three Gaussian distributions, which are centered at 529.8 ± 0.2 eV, 531.1 ± 0.3 eV and 531.9 ± 0.2 eV [31,33,34]. The three peaks denote oxygen in the lattice (M-O), oxygen vacancies (V_O), and chemisorbed oxygen (M-OH), respectively. The carriers in oxide semiconductors mainly originate from oxygen vacancies. From Figure 6, it can be seen that with increasing annealing temperature, the percentage of V_O peak area decreases from 23.45% to 17.36%, while the percentage of M-O peak area increases from 71.15% to 80.47%. This indicates that thermal annealing provides energy for the film rearrangement, enhances the M-O bonding network of films and inhibits the formation of oxygen vacancies, which may lead to the decrease in the optical band gap [35,36]. Meanwhile, the percentage of M-OH peak area also drops to 2.18% when the temperature reaches 300 °C, due to high-temperature annealing reducing the suspension bonds and hydrophilic groups on the film surface.

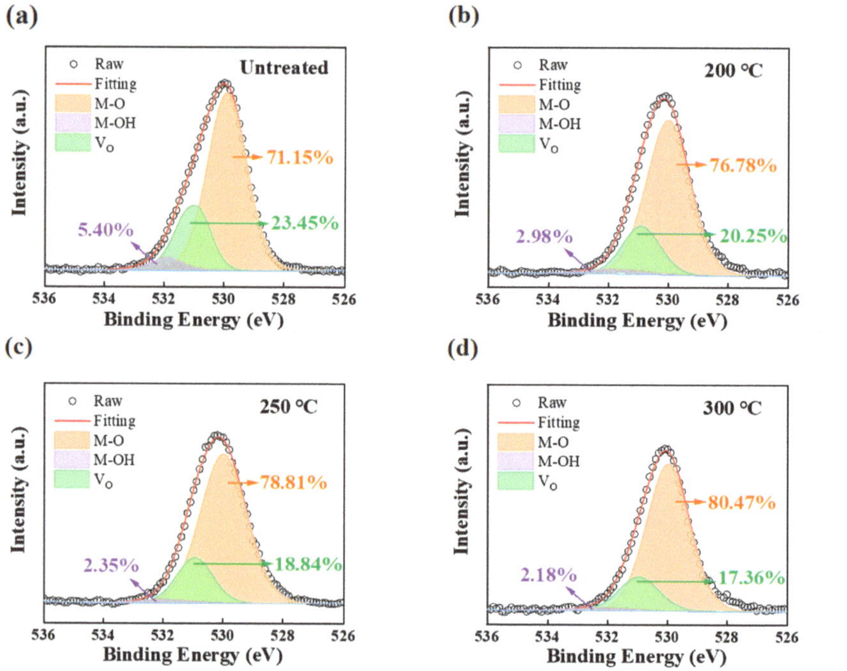

Figure 6. O 1s peak of XPS spectra of PrIZO thin films: (**a**) untreated, (**b**) 200 °C, (**c**) 250 °C, (**d**) 300 °C.

3.6. μ-PCD Analysis

The μ-PCD measurement system is a non-contact and non-destructive test technique, which uses pulsed laser to irradiate the surface of the sample to generate excess photo-generated carriers, resulting in a change of the microwave reflectivity [37]. By observing the temporal response of the microwave reflectivity, it enables the characterization of the capture, recombination, and relaxation of photo-generated carriers so that the shallow-level and deep-level defects in the film can be qualitatively detected [38,39]. From this test, we can obtain peak values and Tau2 values, which characterize deep-level and shallow-level defects, respectively. A higher peak value and a lower Tau2 value mean fewer defects and a better quality of the PrIZO film. Figure 7 shows the μ-PCD results of PrIZO films annealed at different temperature through a mapping scan of the peak level.

As illustrated in Figure 7, it is observed that the peak value of annealed films is higher than that of the untreated film. Since the peak value is associated with deep-level defects, this suggest that annealing has a beneficial effect on the repair of deep-level defects. In addition, with the rise in temperature, the peak value of the film initially increases and subsequently decreases, reaching its maximum at 250 °C. This indicates that the 250 °C-annealed film has the least number of deep-level defects. Moreover, the extracted peak values and Tau2 values varying with annealing temperature are shown in Table 1.

The Tau2 value serves as an indicator of the density of shallow-level defects. As shown in Table 1, the Tau2 value for the untreated film is measured at 2.18 μs. And this value remains unchanged only while annealing at 250 °C. However, it significantly increases at both 200 °C and 300 °C, indicating a corresponding rise in shallow-level defects, which may be detrimental to the performance of the device. From the μ-PCD test, the 250 °C-annealed film obtains the best quality.

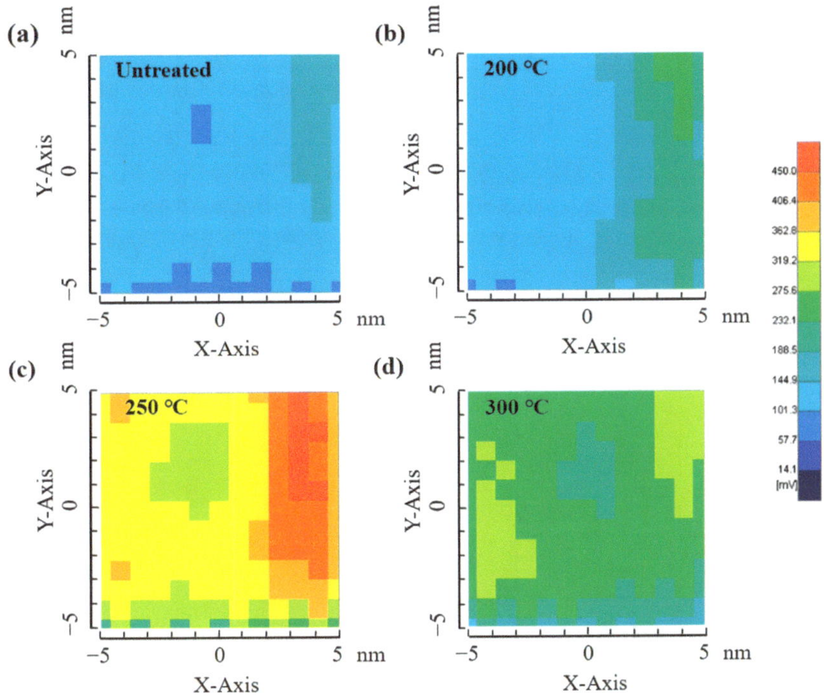

Figure 7. The μ-PCD mapping scan result of PrIZO films: (**a**) untreated, (**b**) 200 °C, (**c**) 250 °C, (**d**) 300 °C.

Table 1. The extracted Peak values and Tau2 values of PrIZO thin films.

Annealing Progress	Peak (mV)	Tau2 (μs)
Untreated	112.9	2.18
200 °C in air	147.7	3.10
250 °C in air	347.6	2.17
300 °C in air	245.2	3.97

3.7. TFT Performance

Figure 8a–c show the output characteristic of PrIZO TFTs. (The electrical characteristic of the device without annealing treatment cannot be measured.) The devices annealed at 200 °C and 250 °C exhibit a clear pinch-off and good current-saturation behaviors. However, the 300 °C-annealed device has a large drain current (I_D) even if the gate voltage (V_G) is not applied. Figure 8d,e show the transfer characteristic of PrIZO TFTs, as obtained through scanning V_G from −20 V to 20 V and recording the corresponding alteration in I_D (V_D = 20 V) and the absolute value of the gate leakage current ($|I_G|$). The saturation mobility (μ_{sat}) and threshold voltage (V_{th}) of devices can be extracted from Equation (2). And the subthreshold swing (SS) of devices can be extracted from Equation (3), so that the interface trap density (D_{it}) can be extracted from Equation (4). In addition, the on currents (I_{on}) and off currents (I_{off}) can be obtained from the transfer curves to calculate the ratio. The extracted performance parameters of PrIZO TFTs are summarized in Table 2. (The V_{th} of the device annealed at 300 °C cannot be extracted in the scanning range.)

$$I_D = \frac{WC}{2L}\mu_{sat}(V_G - V_{th})^2 \quad (2)$$

where W, L, and C are the channel width, length, and gate-insulator capacitance, respectively.

$$SS = \frac{dV_G}{d(\log I_D)} \quad (3)$$

$$D_{it} = \left(\frac{SS\log e}{kT/q} - 1\right)\frac{C}{q^2} \quad (4)$$

where k is the Boltzmann's constant, e is the base of the natural logarithm, T is the temperature, q is the electron charge and C is the gate-insulator capacitance [40].

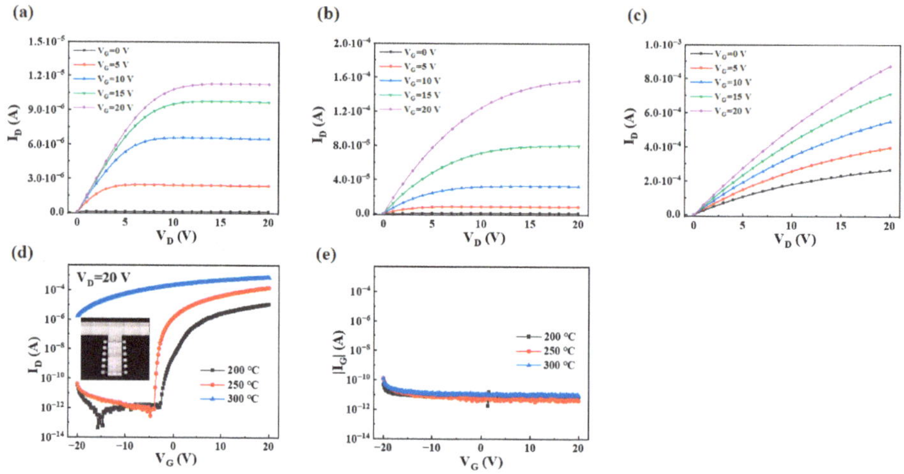

Figure 8. The output curves of PrIZO TFTs: (**a**) 200 °C, (**b**) 250 °C, (**c**) 300 °C and transfer curves of PrIZO TFTs: (**d**) I_D, (**e**) $|I_G|$.

Table 2. Performance parameters of PrIZO TFTs with different annealing temperature.

Temperature	μ_{sat} ($cm^2 \cdot V^{-1} \cdot s^{-1}$)	SS ($V \cdot dec^{-1}$)	D_{it} ($cm^{-2} \cdot eV^{-1}$)	I_{on}/I_{off}	Vth (V)
200 °C	2.80 ± 0.26	0.34 ± 0.04	(1.11 ± 0.16) × 10^{12}	(1.10 ± 0.28) × 10^7	1.35 ± 0.12
250 °C	14.26 ± 0.31	0.14 ± 0.02	(3.17 ± 0.79) × 10^{11}	(1.83 ± 0.35) × 10^8	−1.15 ± 0.14
300 °C	/	2.98 ± 0.12	(1.16 ± 0.05) × 10^{13}	(4.17 ± 0.65) × 10^2	/

As illustrated in Figure 8d, the devices subjected to annealing at 300 °C demonstrate a loss of their switching characteristics, exhibiting instead the characteristics of conductors. On the one hand, high-temperature annealing significantly enhances the density of the semiconductor layer, which is conducive to electron transport, and the reduction in the roughness of the film also reduces the scattering of carriers. On the other hand, the film annealed at 300 °C exhibits the narrowest band gap, which facilitates electron transition from valence band to conduction band. Moreover excessively high temperature may lead to hydrogen diffusion from the dielectric layer into the channel layer, where it acts as a shallow donor, thereby resulting in high-density subgap states [41].

As can be observed in Table 2, as the annealing temperature increases from 200 °C to 250 °C, the mobility of the device increases significantly, and the threshold voltage drifts negatively, due to the changes in density, roughness and band gap of the PrIZO thin films mentioned above. Moreover, the film annealed at 250 °C has a larger peak value and a smaller Tau2 value in the μ-PCD test, which means fewer defects. The corresponding device shows better performance. As a result, the optimized 250 °C-annealed PrIZO TFT exhibits good electric characteristic, with a μ_{sat} of 14.26 $cm^2 \cdot V^{-1} \cdot s^{-1}$, an SS of 0.14 $V \cdot dec^{-1}$, a D_{it} of 3.17×10^{11}, an I_{on}/I_{off} ratio of 1.83×10^8 and a Vth of −1.15 V.

The electrical stabilities of PrIZO TFTs under negative-bias illumination stress (NBIS) were investigated. Figure 9a–c illustrate the changes in the transfer curves of PrIZO TFTs after 3600 s of NBIS under conditions of 250-lux-intensity illumination and −20 V bias voltage. The shift in threshold voltage (ΔV_{th}) was measured at 0 s, 900 s, 1800 s, 2700 s and 3600 s, as shown in Figure 9d. (As mentioned above, the ΔV_{th} of the 300 °C-annealed device cannot be extracted.) As can be seen, with the rise in temperature, the NBIS stability of devices becomes better. Under white-light illumination, the oxygen vacancy above the valence band maximum (VBM) can be ionized to V_O^{2+}, thereby releasing two electrons into the conduction band. At the same time, V_O^{2+} ions are trapped at the insulator/semiconductor interface, which produces a shielding effect on gate voltage [13]. The semiconductor layer annealed at 250 °C exhibits a lower oxygen-vacancy content than that annealed at 200 °C, so the shielding effect is weaker and the ΔV_{th} of the device is smaller during the NBIS test.

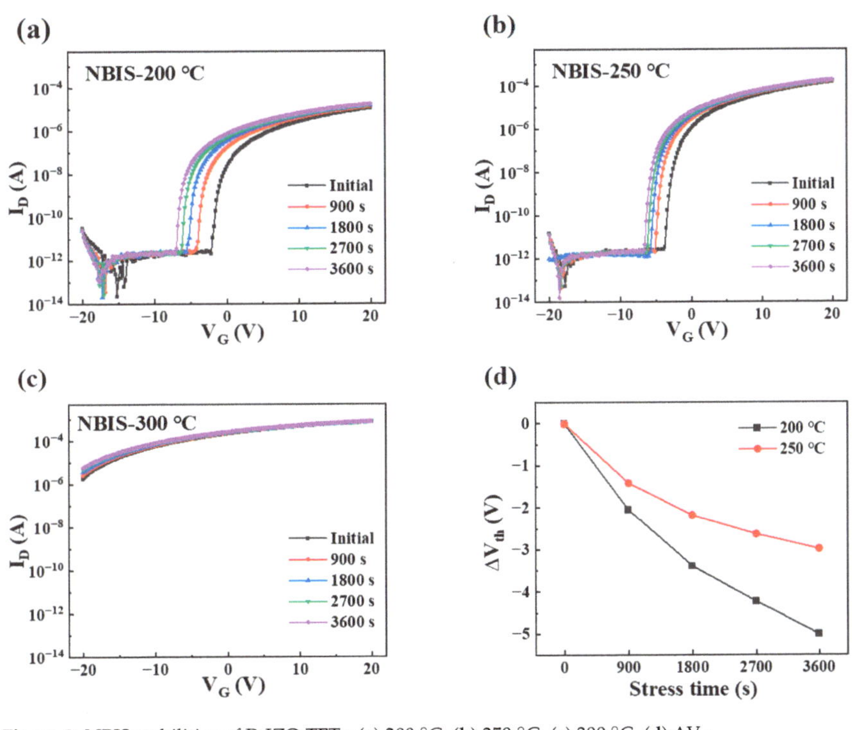

Figure 9. NBIS stabilities of PrIZO TFTs: (**a**) 200 °C, (**b**) 250 °C, (**c**) 300 °C, (**d**) ΔV_{th}.

The flexible PrIZO TFT was prepared on the PI substrate using the optimized annealing process, and its transfer characteristics under different bending radii were measured, as shown in Figure 10. The extracted performance parameters were summarized in Table 3. As illustrated in Table 3, the mobility of the device fabricated on the PI substrate exhibits a slight decrease, accompanied by a positive shift in the threshold voltage. This phenomenon may be attributed to the different surface roughness of insulating layers mentioned in the AFM test.

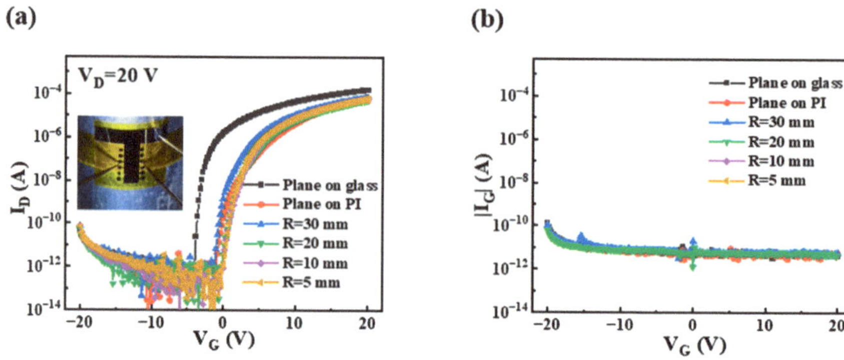

Figure 10. The transfer curves of flexible PrIZO TFTs: (**a**) I_D, (**b**) $|I_G|$.

Table 3. Performance parameters of flexible PrIZO TFTs.

Radius	μ_{sat} ($cm^2 \cdot V^{-1} \cdot s^{-1}$)	SS ($V \cdot dec^{-1}$)	D_{it} ($cm^{-2} \cdot eV^{-1}$)	I_{on}/I_{off}	Vth (V)
Plane on glass	14.26 ± 0.31	0.14 ± 0.02	$(3.17 \pm 0.79) \times 10^{11}$	$(1.83 \pm 0.35) \times 10^{8}$	−1.15 ± 0.14
Plane on PI	12.48 ± 0.40	0.22 ± 0.04	$(6.34 \pm 1.58) \times 10^{11}$	$(4.45 \pm 1.56) \times 10^{7}$	5.54 ± 0.56
R = 30 mm	11.23 ± 0.30	0.26 ± 0.03	$(7.93 \pm 1.19) \times 10^{11}$	$(5.66 \pm 1.73) \times 10^{7}$	3.43 ± 0.33
R = 20 mm	10.86 ± 0.33	0.25 ± 0.04	$(7.53 \pm 1.58) \times 10^{11}$	$(5.37 \pm 1.33) \times 10^{7}$	5.04 ± 0.41
R = 10 mm	10.85 ± 0.25	0.24 ± 0.03	$(7.13 \pm 1.19) \times 10^{11}$	$(6.78 \pm 2.12) \times 10^{7}$	4.50 ± 0.32
R = 5 mm	10.87 ± 0.28	0.21 ± 0.05	$(5.95 \pm 1.98) \times 10^{11}$	$(5.29 \pm 1.08) \times 10^{7}$	4.57 ± 0.29

When the device is mechanically bent, it is observed that there is a slight decrease in the performance of the device, which is attributed to the fact that when mechanical stress is applied, the source/drain electrode and active layer interface may act as the stress concentration points, and these points may create charge trapping or defects [42]. However, even measured at a bending radius of just 5 mm, the PrIZO flexible device continues to demonstrate excellent performance, with a μ_{sat} of 10.87 $cm^2 \cdot V^{-1} \cdot s^{-1}$, an SS of 0.21 $V \cdot dec^{-1}$, a D_{it} of 5.95×10^{11}, an I_{on}/I_{off} ratio of 5.29×10^7, and a Vth of 4.57 V. This is also proof of the great potential of PrIZO in the field of flexible electronics.

4. Conclusions

In summary, the characteristics of PrIZO films and TFTs under various annealing temperatures were investigated. From the μ-PCD measurement, the film with 250 °C-annealing treatment achieves the best quality among all the films. The high conductivity of the 300 °C annealed device may be caused by the high density, low roughness and narrow band gap of the channel layer. The optimized 250 °C -annealed PrIZO TFT demonstrates outstanding performance, with a mobility of up to 14.26 $cm^2 \cdot V^{-1} \cdot s^{-1}$ and a ΔV_{th} of −3 V in the NBIS stability test. When the optimized annealing process is employed in the preparation of flexible TFT, the flexible device still maintains good performance, and the mobility of the device is only reduced by approximately 10% under the bending radius of 5mm. The excellent performance of the device indicates its considerable application potential in transparent and flexible displays.

Author Contributions: Conceptualization, Z.W. and H.N.; methodology, Z.W.; software, H.L.; validation, D.L., R.Y. and X.W.; formal analysis, Z.L.; investigation, G.S.; resources, H.N.; data curation, Z.W.; writing—original draft preparation, Z.W.; writing—review and editing, H.N.; visualization, D.Y.; supervision, R.Y.; project administration, J.P.; funding acquisition, D.L. All authors have read and agreed to the published version of the manuscript.

Funding: This research was funded by the National Key R&D Program of China (No. 2021YFB3600604), the National Natural Science Foundation of China (Grant No. 62174057, 62375057, 62074060 and 22090024), the Guangdong Natural Science Foundation (No. 2024A1515012216 and 2023A1515011026), the Educational Commission of Guangdong Province (Grant Nos. 2022ZDZX1002), the State Key Lab of Luminescent Materials and Devices (Skllmd-2024-05), the Southwest Institute of Technology and Engineering Cooperation Fund (HDHDW59A020301), the Science and Technology Program of Guangdong (Grant No. 2024A0505040026), the Guangdong Basic and Applied Basic Research Foundation (Grant No. 2022A1515140064, 2023B1515120046 and 2024A1515012019), the Science and Technology Program of Guangzhou (Grant No. 2023A03J0024), the Research Project of Guangzhou University (Grant No. ZH2023006), the CUI CAN Program of Guangdong Province (CC/XM-202401ZJ0201), the Guangdong Provincial Key Laboratory of Optical Information Materials and Technology (2023B1212060065), and the Key Laboratory of Optoelectronic Devices and Systems of the Ministry of Education and Guangdong Province.

Data Availability Statement: Data are contained within the article.

Conflicts of Interest: The authors declare no conflicts of interest.

References

1. Kim, D.; Kim, Y.; Lee, S.; Kang, M.S.; Kim, D.H.; Lee, H. High Resolution a-IGZO TFT Pixel Circuit for Compensating Threshold Voltage Shifts and OLED Degradations. *IEEE J. Electron Devices Soc.* **2017**, *5*, 372–377. [CrossRef]
2. Xu, W.Y.; Li, H.; Xu, J.B.; Wang, L. Recent Advances of Solution-Processed Metal Oxide Thin-Film Transistors. *ACS Appl. Mater. Interfaces* **2018**, *10*, 25878–25901. [CrossRef]
3. Natu, K.; Laad, M.; Ghule, B.; Shalu, A. Transparent and flexible zinc oxide-based thin-film diodes and thin-film transistors: A review. *J. Appl. Phys.* **2023**, *134*, 27. [CrossRef]
4. Zhang, G.M.; Xu, Y.C.; Haider, M.; Sun, J.; Zhang, D.K.; Yang, J.L. Printing flexible thin-film transistors. *Appl. Phys. Rev.* **2023**, *10*, 44. [CrossRef]
5. Panca, A.; Panidi, J.; Faber, H.; Stathopoulos, S.; Anthopoulos, T.D.; Prodromakis, T. Flexible Oxide Thin Film Transistors, Memristors, and Their Integration. *Adv. Funct. Mater.* **2023**, *33*, 24. [CrossRef]
6. Park, J.S.; Maeng, W.J.; Kim, H.S.; Park, J.S. Review of recent developments in amorphous oxide semiconductor thin-film transistor devices. *Thin Solid Films* **2012**, *520*, 1679–1693. [CrossRef]
7. Yu, X.G.; Marks, T.J.; Facchetti, A. Metal oxides for optoelectronic applications. *Nat. Mater.* **2016**, *15*, 383–396. [CrossRef] [PubMed]
8. Hu, K.; Guo, Z.; Wang, J.W.; Lu, C.Y.; Wang, M.L.; Wang, T.Y.; Liao, F.X.; Yang, G.H.; Lu, N.D.; Li, L. Tri-Layer Heterostructure Channel of a-IGZO/a-ITZO/a-IGZO Toward Enhancement of Transport and Reliability in Amorphous Oxide Semiconductor Thin Film Transistors. *Adv. Electron. Mater.* **2024**, *9*, 2400266. [CrossRef]
9. Kim, J.; Bang, J.; Nakamura, N.; Hosono, H. Ultra-wide bandgap amorphous oxide semiconductors for NBIS-free thin-film transistors. *APL Mater.* **2019**, *7*, 4. [CrossRef]
10. Wager, J.F.; Yeh, B.; Hoffman, R.L.; Keszler, D.A. An amorphous oxide semiconductor thin-film transistor route to oxide electronics. *Curr. Opin. Solid State Mater. Sci.* **2014**, *18*, 53–61. [CrossRef]
11. Li, Y.; Zeng, X.; Ye, Q.; Yao, R.; Zhong, J.; Fu, X.; Yang, Y.; Li, M.; Ning, H.; Peng, J. Effect of oxygen defect on the performance of Nd: InZnO high mobility thin-film transistors. *Surf. Interfaces* **2022**, *33*, 102184. [CrossRef]
12. Li, Y.; Yao, R.; Zhong, J.; Yang, Y.; Liang, Z.; Fu, Y.; Zeng, X.; Su, G.; Ning, H.; Peng, J. The Hump Phenomenon and Instability of Oxide TFT Were Eliminated by Interfacial Passivation and UV + Thermal Annealing Treatment. *ACS Appl. Electron. Mater.* **2023**, *5*, 4846–4862. [CrossRef]
13. Han, Y.; Chen, Y.; Li, M.; Xu, H.; Xu, M.; Wang, L.; Peng, J. Abnormal Positive Shift of Threshold Voltage in Praseodymium-Doped InZnO-TFTs Under Negative Bias Illumination Temperature Stress. *IEEE Trans. Electron Devices* **2024**, *71*, 1951–1956. [CrossRef]
14. He, P.; Xu, H.; Lan, L.; Deng, C.; Wu, Y.; Lin, Y.; Chen, S.; Ding, C.; Li, X.; Xu, M.; et al. The effect of charge transfer transition on the photostability of lanthanide-doped indium oxide thin-film transistors. *Commun. Mater.* **2021**, *2*, 86. [CrossRef]
15. Kim, I.H.; Kim, S.J.; Kim, S.-J.; An, T.K.; Jeong, Y.J. The impact of nickel doping on metal-oxide network in solution-processed indium zinc oxide transistors. *Mater. Today Commun.* **2023**, *35*, 106221. [CrossRef]

16. Tarsoly, G.; Lee, J.-Y.; Heo, K.-J.; Kim, S.-J. Doping of Indium Oxide Semiconductor Film Prepared Using an Environmentally Friendly Aqueous Solution Process with Sub-1% Molybdenum to Improve Device Performance and Stability. *ACS Appl. Electron. Mater.* **2023**, *5*, 4308–4315. [CrossRef]
17. Xu, H.; Xu, M.; Li, M.; Chen, Z.; Zou, J.; Wu, W.; Qiao, X.; Tao, H.; Wang, L.; Ning, H.; et al. Trap-Assisted Enhanced Bias Illumination Stability of Oxide Thin Film Transistor by Praseodymium Doping. *ACS Appl. Mater. Interfaces* **2019**, *11*, 5232–5239. [CrossRef] [PubMed]
18. Zhang, H.; Liang, L.; Wang, X.; Wu, Z.; Cao, H. Praseodymium-Doped In-Sn-Zn-O TFTs With Effective Improvement of Negative-Bias Illumination Stress Stability. *IEEE Trans. Electron Devices* **2022**, *69*, 152–155. [CrossRef]
19. Zhang, K.; Yao, R.; Fu, X.; Cai, W.; Li, Y.; Xu, W.; Wu, Z.; Luo, C.; Ning, H.; Peng, J. UV irradiation assisted low-temperature process for thin film transistor performance improvement of praseodymium-doped indium zinc oxide. *J. Phys. D Appl. Phys.* **2023**, *57*, 7. [CrossRef]
20. Zou, W.; Liang, Z.; Fu, X.; Ning, H.; Zeng, X.; Fu, Y.; Wang, H.; Luo, C.; Yao, R.; Peng, J. Improvement of PrIZO Thin Films by O2 Plasma Treatment Combined With Low-Temperature Annealing for Thin-Film Transistors. *IEEE Trans. Electron Devices* **2023**, *70*, 5672–5677. [CrossRef]
21. Ide, K.; Nomura, K.; Hosono, H.; Kamiya, T. Electronic Defects in Amorphous Oxide Semiconductors: A Review. *Phys. Status Solidi A-Appl. Mat.* **2019**, *216*, 1800372. [CrossRef]
22. Yuqing, Z.; Zhihe, X.; Jiapeng, L.; Yang, S.; Sisi, W.; Lei, L.; Shengdong, Z.; Hoi-Sing, K.; Man, W. Systematic Defect Manipulation in Metal Oxide Semiconductors towards High-Performance Thin-Film Transistors. In Proceedings of the 2020 4th IEEE Electron Devices Technology & Manufacturing Conference (EDTM), Penang, Malaysia, 6–21 April 2020. [CrossRef]
23. Lee, H.; Jyothi, C.; Baang, S.; Kwon, J.H.; Bae, J.H. Influence of structural defects in solution-processed InZnO semiconductors on the electrical stability of thin-film transistors. *J. Korean Phys. Soc.* **2016**, *69*, 1688–1693. [CrossRef]
24. Pi, T.T.; Xiao, D.Q.; Yang, H.; He, G.; Wu, X.H.; Liu, W.J.; Zhang, D.W.; Ding, S.J. High-Performance a-IGZO TFT Fabricated With Ultralow Thermal Budget via Microwave Annealing. *IEEE Trans. Electron Devices* **2022**, *69*, 156–159. [CrossRef]
25. Kang, C.M.; Kim, H.; Oh, Y.W.; Baek, K.H.; Do, L.M. Pre-Annealing Effect for Low-Temperature, Solution-Processed Indium Oxide Thin-Film Transistors. *J. Nanosci. Nanotechnol.* **2017**, *17*, 3293–3297. [CrossRef]
26. Tarsoly, G.; Lee, J.Y.; Kim, S.J. Improving the photoswitching performance of a transistor with amorphous metal oxide semiconductor thin film by a gradient annealing approach. *Opt. Mater.* **2024**, *157*, 6. [CrossRef]
27. Rim, Y.S.; Jeong, W.; Ahn, B.D.; Kim, H.J. Defect reduction in photon-accelerated negative bias instability of InGaZnO thin-film transistors by high-pressure water vapor annealing. *Appl. Phys. Lett.* **2013**, *102*, 4. [CrossRef]
28. Park, M.; Yoo, J.; Lee, H.; Song, H.; Kim, S.; Lim, S.; Park, S.; Jeong, J.H.; Kim, B.; Lee, K.; et al. Rapid Thermal Annealing under O_2 Ambient to Recover the Deterioration by Gamma-Ray Irradiation in a-IGZO TFTs. *Electron. Mater. Lett.* **2024**, *10*, 1–8. [CrossRef]
29. Xiao, R.B.; Cheng, J.; Lu, Z.Y.; Sun, Q.; Wang, X.; Fu, X.Y.; Gao, J.N. Impact of In-doping and post-annealing on the properties of SnO_2 thin films deposited by magnetron sputtering. *Phys. Scr.* **2024**, *99*, 095937. [CrossRef]
30. Li, M.; Lan, L.F.; Xu, M.; Xu, H.; Luo, D.X.; Xiao, P.; Peng, J.B. Performance improvement of oxide thin-film transistors with a two-step-annealing method. *Solid-State Electron.* **2014**, *91*, 9–12. [CrossRef]
31. Lu, K.; Yao, R.; Wang, Y.; Ning, H.; Guo, D.; Liu, X.; Tao, R.; Xu, M.; Wang, L.; Peng, J. Effects of praseodymium doping on the electrical properties and aging effect of InZnO thin-film transistor. *J. Mater. Sci.* **2019**, *54*, 14778–14786. [CrossRef]
32. Zhang, C.H.; Chen, Y.P.; Hou, C.Y.; Wang, G.; Zhang, Q.H.; Li, Y.G.; Wang, H.Z. Thermal-assisted brush printing of water-based In-Ga-Zn oxide transistors. *J. Alloys Compd.* **2021**, *862*, 158001. [CrossRef]
33. Tang, H.; Lu, K.; Xu, Z.; Ning, H.; Yao, D.; Fu, X.; Yang, H.; Luo, D.; Yao, R.; Peng, J. Effect of Sputtering Oxygen Partial Pressure on the Praseodymium-Doped InZnO Thin Film Transistor Using Microwave Photoconductivity Decay Method. *Micromachines* **2021**, *12*, 1044. [CrossRef] [PubMed]
34. Hu, M.Z.; Xu, L.; Zhang, X.N.; Hao, H.Y.; Zong, S.; Chen, H.M.; Song, Z.C.; Luo, S.J.; Zhu, Z.H. High mobility amorphous InSnO thin film transistors via low-temperature annealing. *Appl. Phys. Lett.* **2023**, *122*, 033503. [CrossRef]
35. Peng, J.W.; Liu, P.C.; Lee, S. Reversible band gap tuning of metal oxide films using hydrogen and oxygen plasmas. *Thin Solid Films* **2013**, *531*, 81–87. [CrossRef]
36. Ali, A.V.M.; Kekuda, D. Thickness and oxygen partial pressure dependence on optical band gap of indium oxide by reactive evaporation method. *IOP Conf. Series: Mater. Sci. Eng.* **2015**, *73*, 012027. [CrossRef]
37. Yasuno, S.; Kita, T.; Morita, S.; Kugimiya, T.; Hayashi, K.; Sumie, S. Transient photoconductivity responses in amorphous In-Ga-Zn-O films. *J. Appl. Phys.* **2012**, *112*, 053715. [CrossRef]
38. Chen, J.; Hu, S.; Ning, H.; Fang, Z.; Tao, R.; Zhou, Y.; Cai, W.; Liu, X.; Yao, R.; Peng, J. Evaluation of Nd–Al doped indium-zinc oxide thin-film transistors by a μ-PCD method. *Semicond. Sci. Technol.* **2019**, *34*, 055011. [CrossRef]
39. Goto, H.; Tao, H.; Morita, S.; Takanashi, Y.; Hino, A.; Kishi, T.; Ochi, M.; Hayashi, K.; Kugimiya, T. In-line Process Monitoring for Amorphous Oxide Semiconductor TFT Fabrication using Microwave-detected Photoconductivity Decay Technique. *IEICE Trans. Electron.* **2014**, *E97.C*, 1055–1062. [CrossRef]

40. Cai, W.S.; Park, S.; Zhang, J.W.; Wilson, J.; Li, Y.P.; Xin, Q.; Majewski, L.; Song, A.M. One-Volt IGZO Thin-Film Transistors With Ultra-Thin, Solution-Processed Al_xO_y Gate Dielectric. *IEEE Electron Device Lett.* **2018**, *39*, 375–378. [CrossRef]
41. Bang, J.; Matsuishi, S.; Hosono, H. Hydrogen anion and subgap states in amorphous In–Ga–Zn–O thin films for TFT applications. *Appl. Phys. Lett.* **2017**, *110*, 232105. [CrossRef]
42. Sheng, J.; Jeong, H.-J.; Han, K.L.; Hong, T.H.; Park, J.S. Review of recent advances in flexible oxide semiconductor thin-film transistors. *J. Inf. Disp.* **2017**, *18*, 159–172. [CrossRef]

Disclaimer/Publisher's Note: The statements, opinions and data contained in all publications are solely those of the individual author(s) and contributor(s) and not of MDPI and/or the editor(s). MDPI and/or the editor(s) disclaim responsibility for any injury to people or property resulting from any ideas, methods, instructions or products referred to in the content.

MDPI AG
Grosspeteranlage 5
4052 Basel
Switzerland
Tel.: +41 61 683 77 34

Micromachines Editorial Office
E-mail: micromachines@mdpi.com
www.mdpi.com/journal/micromachines

Disclaimer/Publisher's Note: The title and front matter of this reprint are at the discretion of the Guest Editor. The publisher is not responsible for their content or any associated concerns. The statements, opinions and data contained in all individual articles are solely those of the individual Editor and contributors and not of MDPI. MDPI disclaims responsibility for any injury to people or property resulting from any ideas, methods, instructions or products referred to in the content.

www.ingramcontent.com/pod-product-compliance
Lightning Source LLC
Chambersburg PA
CBHW041101270926
51850CB00025B/2946
* 9 7 8 3 7 2 5 8 3 3 0 0 9 *